KNOWING the OUTDOORS in the DARK

KNOWING
the
OUTDOORS
in the DARK

VINSON BROWN

with many illustrations by
Phyllis Thompson

Stackpole Books

KNOWING THE OUTDOORS IN THE DARK
Copyright © 1972 by
THE STACKPOLE COMPANY
Published by
STACKPOLE BOOKS
Cameron and Kelker Streets
Harrisburg, Pa. 17105

Printed in U.S.A.

Library of Congress Cataloging in Publication Data

Brown, Vinson, 1912-
    Knowing the outdoors in the dark.

    Bibliography: p.
    1. Natural history--Outdoor books. 2. Outdoor
life--Safety measures. I. Title.
QH81.B856   574   71-179605
ISBN 0-8117-0933-7

To my daughter Roxana,
whose devotion to beautiful music,
both played herself on the piano and
from stereophonic records, is an
inspiration to me in writing about
the music and nature of the wilderness

Contents

Common insects and relatives attracted to lights
Common insects attracted to sweet formulae
Insects coming to special flowers
Large insects and other invertebrates crawling over ground, particularly in desert areas
Insects and their relatives swimming in pools or walking on bottom
Insects and their relatives crawling on tree trunks and steep rock surfaces
Insects coming to scent of female from long distance
Insects that flash lights in darkness

7   Knowing Life by the Waterside   105

Zones of life on the seashore
How seashore organisms protect themselves
Mysteries of the seashore night
   • how the sunflower starfish overpowers a mussel • rainbow-hued colors of life in a low-tide cave
Guide to sights of:
   • fish of the tide pools or the sands • lobsters, crabs, shrimps, etc. • seashells, snails, etc. • bivalves • starfish, brittle stars, sea urchins, sand dollars, sea cucumbers • segmented worms • jellyfish, hydroids, sea pens, sea anemones, etc.
Wildlife dramas by a pond in the woods
   • life-and-death struggle between muskrat and mink • mating dance of the mayflies • antics of the fun-loving otters • what happens when a raccoon family meets a skunk family • how the horned owl cuts short the moonlight merriment of cottontail rabbits

8   Examining Plant Life After Dark   117

How plants move in the wind
Nighttime personalities of trees and forests
Plants as shelter for animals and birds
Identification guide to:
   • molds and fungi • gymnosperms or coniferous trees • monocotyledonous or parallel-veined plants • dicotyledonous or net-veined plants • flowers that open their petals mainly at night

9   Listening In on Nature's Nighttime Sounds   123

The sound of wind and waves at night
Communications systems of animals
   • how birds and animals stake out their territorial claims • how animals warn each other of danger • breeding calls of frogs and toads
Guide to bird, insect, and animal sounds
   • bird calls, notes, songs, and hoots • musical bird and amphibian calls • nonmusical sounds of birds or insects, etc. • sounds of mammals • musical choruses of amphibians and birds • nonmusical amphibian choruses • musical insect choruses

10   Smelling Out Nighttime Odors   154

Techniques for sharpening the sense of smell
Guide to some common nighttime odors
   • sweet • pleasant but not sweet or musky • musky • unpleasant but not musky

Acknowledgments

I WISH TO EXPRESS my deep appreciation to the following people for their kind assistance with this book: to Mrs. Florence Musgrave, my assistant and secretary, for many hours of typing spent getting the manuscript, a very complicated one, into shape; to the editors at Stackpole Books for their patience and careful explanations of what was needed until we reached a workable book; to the following artists who contributed to the book: Miss Phyllis Thompson, who did a fine job contributing the great majority of the art; Sara Thompson; Douglas Andrews; Jerry Buzzell, James Gordon Irving, Iain Baxter, Emily Reid, Rune Hapness, Carol Lyness, Gene M. Christman, Constance Gabriel Schulte, Don G. Kelley, Jean Colton, and Ben Cummings for their fine art work reproduced from the following Naturegraph books: *Wildlife and Plants of the Cascades, California Wildlife Map Book, The Sierra Nevadan Wildlife Region, The Californian Wildlife Region, Exploring Pacific Coast Tide Pools, Common Seashore Life of Southern California,* and *Common Seashore Life of the Pacific Northwest;* and William Carey Grimm for drawings reproduced from his *The Book of Trees,* a Stackpole book.

Many thanks also to Houghton Mifflin Company for permission to include in this book geographical range maps for various animals based on maps in the following books: *A Field Guide to the Mammals,* Second Edition, Revised and Enlarged. Copyright © 1952, 1964 by William Henry Burt and Richard Philip Grossenheider. *A Field Guide to Reptiles and Amphibians.* Copyright © 1958 by Roger Conant. *A Field Guide to Western Reptiles and Amphibians.* Copyright © 1966 by Robert C. Stebbins.

Appreciation is likewise extended to Little, Brown and Company, publishers of my *How to Make a Miniature Zoo,* from which several illustrations by Don G. Kelley are reproduced in this book.

I am grateful too for the information gathered in research from all the books listed in the Suggested References, and many other sources.

Introduction

MANY PEOPLE FIND themselves in the dark, in the outdoors at night, for some reason or another. Some have gone for a walk and found the dark coming quicker than they imagined. Others may have a car break down and so be on a dark road for a while. Some, such as woodsmen and messengers, have to go through dark areas as a part of their work. Others may be camping out, and of course, be surrounded by darkness beyond their fire and lamplight. Still others may be vacationing in a cabin on the edge of a woods, lake, desert, or some other wild area. Their cabin or tent may even be in the midst of such a wilderness. Some people go on hikes or walks and get lost in the dark; they usually become thoroughly frightened. Yet, how few of these people realize what a grand and still safe adventure the dark can be, and how filled with interest and wonder! For all these people this book is meant to give knowledge, reassurance, and help in making the darkness of the outdoors not something to look upon with dread or boredom, but with delight and anticipation because of the many things to find and do in the nighttime hours. This world we live in is a dual one, blanketed half with darkness and half with light, and it is a shame to let the dark half go by without realizing its potential. One of the rewards of exploring the outdoors at night is the chance to meet many interesting nocturnal creatures.

Many large museums or zoos have exhibits, even live exhibits, of animal activity at night. If possible, study some of these exhibits and learn the animals, birds, and insects by name before starting to watch night life in the outdoors.

Exploring the Outdoors at Night
In Safety and Comfort

This book provides a number of aids to understanding the dark and making the most of your presence in it, while losing fear and gaining respect for and knowledge of wildlife. Here are some preliminary suggestions:

1. It is wise to move cautiously in the outdoors at night because there may be some dangers (described in Chapter 13), but beware of fear and panic because they may become more dangerous than anything else in the night. This is especially true when a person becomes lost and lets panic drive him to run in circles until he becomes exhausted and a possible victim of exposure. So enter the dark gradually at first, if possible, remembering there is really very little to fear except the foolishness which results from letting fear take the upper hand.

2. Many people have become so accustomed to turning on a light or going to a lighted place when it gets dark, that they draw back from the darkness without giving themselves a chance to know it, thus missing valuable experiences. Reverse this reaction by peering into the darkness with curiosity. It is a fact that human eyes grow in power the longer they stay away from light (see Chapter 1). Start to listen to and feel the night as something worthy of your attention. Soon the night wind and the sounds it wafts may begin to bring pleasant messages.

3. It is best to prepare for exploring the dark by wearing comfortable, warm, dark clothes that nighttime creatures cannot see very well. Try streaking your face with charcoal (easy to wash off later), so the white area does not alarm the animals.

4. Bring a powerful flashlight that can be fastened to your hat or head by a band so as to leave both hands free, but be sure to cover the shining end with red paper or plastic so the light that comes through is red in color. Many night creatures are not alarmed by this red light because they actually do not see it (why will be explained in Chapter 1).

5. A pair of binoculars is helpful for increasing far vision, and a compass makes it possible to proceed in a straight line if there is any danger of getting lost.

6. High-topped hiking shoes or boots, well-greased against the wet, and good jeans give protection against bites from poisonous snakes. A mosquito netting over a hat, or a good mosquito repellant comes in handy in mosquito country. Matches should be in a waterproof box or plastic bag in wet weather if a fire is needed, and a hand axe or hatchet may be necessary to make kindling from semidry sticks found under logs or rocks.

7. To avoid being sensed by wild animals, rub your body and clothes with a strong-smelling bunch of leaves, as from sagebrush in the West, balsam fir needles in the East, or honeysuckle in the South.

8. Before leaving on an extensive trip, tell friends or family exactly where you plan to go so they will know where to look for you if anything goes wrong. Of course, if an experienced person goes along, you are in luck and can gain much knowledge from watching and listening to him. But never plan to learn much from the dark with a group of more than three or four really interested people, as numbers make too much noise and disturbance.

9. Learn to be still and watch, listen, smell, and feel. Standing or sitting still in one place for a long time or even some fairly good pauses while on a hike make it possible to sense far more of the surroundings and what is in them. Movement means not only more noise, but motions that cause alarm to all nearby birds and animals.

10. Don't be ashamed to take something that will make you more comfortable. A piece of foam rubber is excellent for placing on top of a rock or stump being used

as a wilderness chair, or for lying upon on the ground so that you will be able to stay still without being constantly prodded by twigs or rocks. Even a folding camp chair may enable you to sit still for a long time in one place and watch and listen. It is especially useful in a blind (see Chapter 1).

11. Above all, be prepared to be alert and use all your senses, for there is no telling what wonder may suddenly appear in the outdoors at night.

Sharpening the Senses

The chapters that follow describe in more detail how to sharpen each sense. All of the senses, even the little-known sixth sense, can usually be sharpened far beyond their present acuity. The noises, flashing neon signs, heavy smells and other deadening effects of civilization tend to make sight, hearing, smell, and even touch less effective than they should be. The result is that, when people get into the outdoors, they may be completely blind or deaf to interesting things and happenings around them that a more alert and sense-sharpened person would know about immediately.

Those dulled senses can be improved in a comparatively short time with a little practice, interest, and faith. Just as the hand and eye that use a hammer become gradually more expert with use and practice, so any sense can be improved for use in the darkness. To sharpen a sense means to practice its alert use constantly in the outdoors, paying close attention to every sight, sound, smell, or other sensation that comes from the surroundings, until nerves that were once dull and lethargic because of the influence of city life come awake and active again, bringing messages out of the wilderness that make life far more interesting and exciting. Knowing that you can do this is to have a faith that creates miracles!

How to Use This Book

Because man perceives wildlife, vegetation, and terrain with his senses during the hours of darkness as well as during the hours of daylight, this book is organized according to sense impressions transmitted by the senses of sight, hearing, and smell. Since vision is more hampered by darkness than is any other sense, Chapter 1 offers some techniques for improving eyesight in the dark and some guidelines for interpreting movements and objects that glow in the dark. Chapters 2 through 8 describe various creatures and plants that can be seen at night. In most cases, the appearance and habitat of each animal and plant are described, and the geographical range of each animal is shown by a map. Typical activities of nocturnal creatures are also described. Most animals and plants (or groups of animals or plants) listed in Chapters 2 through 8 are illustrated. To find information on the following classes of living things which can be seen at night, consult the chapters as indicated:

Mammals	Chapter 2
Birds	Chapter 3
Reptiles and amphibians	Chapter 4
Fish	Chapter 5
Insects	Chapter 6
Life by the waterside	Chapter 7
Plants	Chapter 8

Chapter 9 deals with the sounds made by wildlife. It explains how the sounds animals make, the movements they make, and the scents they emit combine to form a complete system of communication. The creatures listed in Chapter 9 are arranged according to the tonal qualities of the sounds they make: e.g., musical or nonmusical sounds; whistling sounds; loud, medium, or soft sounds, etc. The sounds are described in detail, and the reader is told in what kinds of habitat the sounds can be heard.

Chapter 10 lists some of the common trees, plants, and animals that can be smelled at night and describes the smells.

Chapters 11 and 12 tell the reader how to develop two faculties which will help him to move around in the dark: a sense of direction and the little understood sixth sense.

Chapter 13 discusses necessary safety precautions for nighttime exploring, tells how to behave when encountering a dangerous animal, and lists items of equipment necessary for safety and comfort.

Chapters 14 and 15 explain systematic, scientific methods of learning about nocturnal wildlife. Chapter 14 discusses techniques for approaching and attracting animals, while Chapter 15 outlines several projects which can be used to test the reactions of animals and explore the mysteries of nature. In fact, some of these projects may even put the reader on the road to solving some of these mysteries.

CHAPTER ONE

Reading Nature's Signs in the Dark

No ONE JUST beginning to explore the wonderful world of darkness can possibly learn right away about all that can be seen in North America between sunset and sunrise. This book concentrates on supplying aids in identifying some of the more common things to see and an understanding of what they mean. It will also prepare for exploring some of the deeper mysteries. The night of the outdoors is filled with many mysteries, any of which could probably keep the observer busy through exciting hours, even from the comfort of a cabin porch. Entering the night world is to be surrounded with the least-known area of life's activity outside of the deep sea. Man, in his craving for light and fear of the dark, has often missed one of life's greatest adventures, exploring the dark.

How to Increase Vision at Night

It is easier to see in the dark than most people think possible. This is because most individuals are constantly using artificial light at night and rarely give their eyes a chance to get used to the dark. Even a moonless night generally has enough light from the stars to help people see enough to move around slowly. It may be hard to believe, but a human being generally has eyes that can see better in the gloom of night than those of a bear, and almost as well as those of a cat, if the human practices enough.

In the first fifteen minutes a person stands in the dark his iris spreads to its widest extent, and so catches much of the light around him, but it takes another half-hour, or three-quarters of an hour altogether, for the retina behind the iris to become fully adjusted to the night and able to make full use of that wide-open iris. Go into a familiar place at night, like the backyard, and patiently practice moving around in the dark. As your eyes gradually accustom themselves to the darkness, it will become easy to see ahead. A guiding stick may provide confidence at first, but soon it will not be needed. Practice also moving with the aid of hearing (see Chapter 9), the "direction sense" (Chapter 11), and the sixth sense (Chapter 12).

After becoming familiar with moving about in the backyard, you will be ready to move about in the night in wilderness areas with some assurance. What at first seem to be frightening dark masses begin to take on familiar meaning, and any fear disappears.

The Light That Helps You See Without Disturbing Wildlife

Essentially, however, practice in seeing in the dark is not necessarily for traveling in it, but mainly to be able to watch and sense the life in one's surroundings without disturbing the wild creatures. Try to catch the real feeling of the night. There is one artificial aid for seeing through the gloom that can be used frequently without disturbing most of the wildlife. This is a red-covered flashlight. Its red beam is not seen by most night creatures because they do not have in their eyes the color-

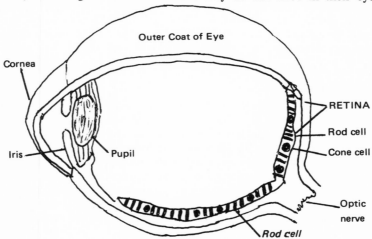

Cross-Section of Mammal's Eye
Because the rod cells in the inner lining outnumber the cone cells, this eye is very sensitive to light in darkness. An eye with more cone cells in the lining would be best for use in daylight.

sensitive cone cells that most daylight-dwelling animals have. Instead they have eyes that are rich mainly in rod cells (see illustration), which are more important for detecting light and shadow, and so are not sensitive to red, which light appears black to them. This makes it possible to put a red plastic screen over a flashlight or even over the headlights of a car and use the red light to explore the darkness without alarming many night-loving animals and birds. Remember, however, that there are some animals, such as foxes and wildcats, that travel about both by daylight and dark, who will notice a red light. But even they will not notice it nearly as much as a white beam of light.

Three Kinds of Inexpensive Blinds
Top: **Teepee made out of several long sticks with green or brown tarp, blanket, or cloth hung over it, with eye holes. Can be put in tree house.** Bottom left: **Large box with view slot hidden with brush.** Bottom right: **Long marsh grass or reeds tied together at top.**

If the flashlight can be fastened to your head, then hands are free while watching the wild creatures. This is particularly useful when using binoculars. Remember to move very slowly and quietly, and to stand or sit still for long periods. It is amazing what magic such prolonged stillness can produce, as many animals and birds of the night simply no longer realize any human presence and begin to act in a normal way.

Blinds and Camouflage

With the use of such a powerful red light, and possibly also good wide-angled binoculars, the world of the night can be explored in safety and comfort from the porch of a cabin. Or a comfortable chair can be placed just outside a tent or in a blind that has been set up in a wild place and covered with mosquito netting or provided with plastic windows. Such blinds are made more effective by disguising them with branches and leaves to look like part of the natural environment. Here again success in seeing interesting happenings and creatures will be helped by sitting perfectly still for long times and watching alertly. Dark clothes, smudging the face with charcoal, and rubbing exposed skin and clothes with a strong-smelling plant that hides the human smell help also.

Using Nature's Own Light Reflectors

When exploring wilderness areas in the dark, you can use natural sources of light wisely and see more. Thus, while there is still light from the day, you can penetrate woods, but it would be wiser not to go into such a dark area without the help of the moonlight unless you take a good flashlight (covered with red plastic). Water areas reflect and increase the light of the stars. Snow catching and intensifying starlight makes it possible to walk through a snowy woods at night with much more ease than usual. Such snow is fascinating, also, because of the many tracks that can often be found on it. It is possible to follow these tracks in the moonlight and read the stories of life and death they tell, and even sometimes come on the animals that left them. This is particularly true if you are walking with snowshoes over wet and deep snow, in which creatures like foxes and deer may flounder and find such hard going that they can be overtaken. If you do, however, do not harm or tease them, but watch with interest and kindness.

Telltale Nighttime Signs

Below are listed some of the common nighttime happenings you are likely to see, with information on how to understand them.

Tree branches waving

What this means. Possible storm coming, but each tree gives its sign by the way it waves and the leaves catch light.

What to do. Identify trees by leaves, bark, flowers, seeds, etc. as shown in a good tree book—such as *The Book of Trees* by William Carey Grimm—but learn the sign of each tree by watching its movements and light patterns under the influence of the wind so you can tell the tree from a distance just by its appearance at such times. This will help in relating the wild animals and birds to the trees with which they are associated. It is also fascinating to see the strangely different ways trees act in the wind. Thus a Douglas fir has branch tips that literally seethe in the wind like octopus tentacles, while a maple shows the dark and light of its leaves in ripples of light and dark that look like running water.

Gleaming points or spots of light

What they may mean. If seen in pairs, they usually mean the eyes of wild creatures reflecting the light of the moon or a flashlight. Flashes of light usually mean fireflies. A row of bright dots means a glowworm or other ground-dwelling creature with phosphorescent light, such as a beetle grub.

What to do. Use a powerful red-covered flashlight to try to see the originator of the flashing eyes or other light. Write down the different colors of light you see and how the colors are associated with different creatures, as shown in the charts below.

IDENTIFICATION KEY TO LIGHTS IN THE NIGHT

Lights flashing on and off through woods	Fireflies (signals between male and female)
Small tiny rows of lights on ground	Glowworms
Bits of light glowing in rotten logs, etc.	Beetle grubs
Bits of light glowing in the sea	Phosphorescent tiny animals, too small to be seen separately

IDENTIFICATION KEY TO EYES IN THE NIGHT

Moving tiny scintillating specks of white	Wolf spiders
Stationary tiny scintillating specks of white	Trapdoor spider
Brilliant ruby red eyes	Alligator or caiman
Close-set red eyes	Black-crowned night heron
Glowing red coals	Woodcock
Large orange red eyes	Yellow-crowned night heron
Small orange red eyes shining on tree trunks	Noctuid moths, particularly Catocala

Closely set, large, bright, orange eyes	Bear
Bright yellow eyes	Raccoon
Yellowish white eyes	Bobcat or Canadian lynx
Dull gleaming white	Whippoorwill
Opalescent green eyes	Bullfrog
Fiery white	Coyotes, dogs, wolves

Faint ghostlike glows on the ground
or on dead trees or logs, sometimes moving very slowly

Meaning. Sometimes made by fungus that grows in rotten wood, or by slime fungus that may cover part of a rotting log and move with a very slow creeping motion over it as if it were an animal instead of a plant.

What to do. Examine closely and watch for any action in these strange plants, noting different colors shown, to see if different kinds can be identified by the various colors of light they have. It is better to watch these in complete darkness without using the red flashlight.

Flash of white tails moving

Meaning. Usually done by cottontail rabbits (low to the ground), or by deer (higher off the ground), as they signal alarm to their friends when they detect danger.

What to do. Stay perfectly still and watch closely. If you are the one who frightened them, they may be back in a fairly short time, for the stillness will calm their fears. If something else has alarmed them, also be still, but watch for faint movement in the dark, which may be caught on the beam of the red flashlight. You may see a fox stalking the rabbits, or a coyote, wildcat, or even a mountain lion stalking a deer. Any danger from a mountain lion is practically nil, as they are anxious to avoid human beings. I have been trailed by mountain lions twice for hours, but their interest in me in the night was purely curiosity.

A small jerking movement, usually off the ground,
as if suspended in air

Meaning. Usually caused by the tail tip of a wildcat, other cat, or mountain lion jerking spasmodically as the cat stalks a mouse, rat, rabbit, deer, etc. It is a sign to other cats that this one claims the prey it is stalking and wants no other to interfere. Sometimes a fox deliberately waves the white tip of its tail to excite the curiosity of a rabbit or other possible victim, while another fox, usually its mate, stalks the prey from behind.

What to do. Be perfectly still and try to catch the whole stalking process in the red flashlight beam, and also watch with binoculars. However, if the carnivore shows nervousness in the light beam, turn it off and try to get your eyes adjusted to the darkness enough to watch. Sometimes, of course, a cat waits in one place for the animal it is preying on to come to it or to come out of a hole. Other times it crouches low and seems to flow slowly over the ground towards its intended victim.

White or silver of water showing in moonlight or starlight
from splashes in pool or stream

Meaning. The flash of silver scales means a fish or several fish jumping for insects, an interesting sight to watch. Water flashing close to the bank is likely to be a raccoon or bear fishing or hunting for fish and crayfish in the shallows, by swift dipping of the paw. If it is a long and fairly large bubbly trail of silver in the water, it is likely to be a beaver, muskrat, mink, otter, or even a merganser duck swimming about. A tinier trail of bubbles could be a water rat or water shrew.

What to do. Approach very cautiously with slow movements in order to get close enough for a good view; then stand or sit perfectly still. Even if the creature has been frightened, if you stay still long enough and are wearing clothes coated with a strong-smelling plant, the animal may come back again and display entertaining antics.

Darting, zigzag movements in the air, especially over or near water

Meaning. These are usually made by bats, if fairly large, or by moths, particularly hawkmoths, if smaller. The bats are seeking to catch insects in the air, while the moths are dodging to escape the bats.

What to do. Sit still and watch. As the evening progresses, notice how different kinds of bats come out at different hours, the pallid bats and pipistrelles (tiny, little mouselike bats) early in the evening, the big brown bats and silver-haired bats later. Try to identify the different kinds from the pictures and descriptions in the Guide to Sights of Mammals at the end of Chapter 2 and note the hours they are active, also their different methods of flying and catching insects. Watch the contest of life and death between the bats and insects, and notice the ways the insects use to escape. Notice how the hawkmoth hovers over flowers. While so hovering, the moth may extend its long tongue as much as a foot to reach down the throat of the flower and find nectar. The tongue is kept coiled under the mouth when not in use. Such moths help fertilize the flowers because of the pollen they carry and are particularly lured by night-loving flowers.

Darting movements through grass

Meaning. These are usually caused by meadow mice and other small creatures, such as shrews and grasshopper mice, running along the meadow mouse trails in the grass and then diving headfirst into a hole at the first sign of danger.

What to do. Sit still and watch to see what kind of animals come. You may even see a least weasel on the blood trail of a mouse, its body slim enough to go down the hole and on even to the mouse nest deep in the ground.

Leaping movements in grass, desert, or semidesert areas

Meaning. Rabbits, of course, jump about in meadows, feeding, playing, or dodging to escape enemies. They may even be seen doing jumping dances on nights of the full moon. Jumping mice also use their long legs to jump about among the grass stems and play. In more desertlike areas such nocturnal leaping is done by the amazing kangaroo rats and mice and their cousins, the pocket mice, who make sudden jumps to escape enemies or leap high to play games or carry on kicking battles with each other high in the air.

What to do. Make yourself comfortable and sit absolutely still. Even if the little creatures jump off to hide, if you are quiet enough they will soon be back with their wonderful circus of jumpers and clowners.

Ghostlike wings bearing thickly feathered bodies
through the air, usually in long-slanted downward glides,
but occasionally more steeply

Meaning. This is the usual attack glide of owls (from the huge great horned owls to the extremely tiny elf owls), as they seek to sink needle-sharp claws into the necks or backs of mice, rats, rabbits and other small creatures of woods or field. They try to come soundlessly, using the shadows to hide in, and seize their prey from behind, before it is aware death is swooping down from above.

What to do. Just be very still and watch the attack. Sometimes the owl misses. When he does, watch his reaction. It may be comical. The owl will probably not be bothered by the red-covered flashlight.

A fluttering circling movement of a dark, plump body up from the ground

Meaning. This flying leap upward is usually done by poorwills, chuck-will's-widows, or whippoorwills, all ghostly gray-brown birds that hide in ground cover or camouflage themselves among the rocks and brush of a desert before springing up swiftly to snap a passing insect. The bill is quite large and flat, with hairs along the side, so it can open wide and form a larger trapping area with the hairs, then easily snap up an insect in flight.

What to do. Move cautiously closer until the whole interesting operation comes into view. It is best seen in the dusk of evening, but these birds are also active in the early dawn.

Small hawk-size birds flying with a dipping, rolling,
back-and-forth motion across the evening sky,
sometimes even diving down almost to the ground
in swooping flight but rarely touching it

Meaning. These are nighthawks, always told by their long narrow wings with white bars near the dark tips. They are catching insects in flight like the bats, but have a different movement and technique. They use their eyes for the job, while bats mainly use sound (see next chapter). Therefore, their movements are not as precise as those of the bats.

What to do. Use a pair of binoculars while sitting on a cabin porch or in an open area to watch the interesting erratic flight of these birds and observe their successes and failures. Sometimes the male shows off to the female, especially at mating time, by diving nearly to the ground and making a loud booming noise.

Vertical or slanted black and white streaks or spots
moving along from about six inches to a foot above the ground

Meaning. These are probably skunks, either the more common and larger cat-size striped skunk, or the rat-size little spotted skunk of the West. They are great hunters of insects, mice, and grubs, which they find by turning over sticks, stones, and other debris, looking beneath them and seizing quickly in mouth or paws whatever they find. They give warning not to come too near by raising their tails, stamping the ground, or even standing upon the front feet like an acrobat, as is done most comically by the spotted skunk. But it is not comical for the intruder when they throw a painfully burning cloud of gas which can reach about ten feet in calm weather or perhaps fifteen feet downwind!

What to do. Don't get closer than about fifteen feet, but have no fear that they can catch you, as skunks are slow-moving animals. Since, because of their strong scent glands, they have little fear of people without guns, it is often possible to follow a skunk for some distance with a red-covered flashlight and observe all its actions. If you move quietly and make no sudden movements, it may get used to you enough to begin to act normally in its hunting.

A slithering, side to side, or coiling motion in the grass or under bushes

Meaning. This is usually a snake or, sometimes, a large lizard or salamander. None of the soft and moist-skinned salamanders are of the slightest danger to a human, unless he eats one of the poisonous type. The only dangerous lizard in the United States is the Gila monster of the Southwest, which minds its own business and is very unlikely to attack people. It is sluggish in movement, and so could never catch you.

Most snakes are completely harmless, but watch out for the pit vipers, such as rattlesnakes (usually told by their rattles), copperheads, and cottonmouth moccasins, all told by their very wide heads (compared to the narrow neck), and the pit between the eye and the nostril. These dark pits are heat-sensing organs, used by the pit vipers in tracking down warm-blooded animals, such as rabbits or rats, or in sensing their approach while lying in wait. The only other dangerous snake is the coral snake of Arizona and New Mexico, which is told by its brilliant red and black bands, always separated by narrower yellow bands. The coral snake is not nearly as aggressive as the pit vipers and may not bite even when picked up, but should not be handled as it is deadly poisonous.

What to do (see Chapter 13 on how to avoid night dangers). Snakes at night are usually hunting for small creatures, and are interesting to follow with a red-light flashlight by moving very quietly and cautiously. Don't get too close and keep alert for the poisonous snakes.

There is one very interesting way to find night snakes in desert or semidesert areas with comparative safety. This is to drive along a black-top road where traffic is light late at night, moving very slowly, and carefully watching the road ahead. Many snakes come to these roads for warmth as the night chill increases towards morning. Notice that each kind of snake has its own characteristic way of acting and moving.

Scuttling and jumping movements on dark beaches at night

Meaning. Ocean beaches are extremely interesting places to visit at night, as almost all forms of life there become extra active. The most obvious movements are those of the beach fleas of various sizes jumping about among the kelp and other seaweed drying on the beach, while down wherever there are clusters of rocks the rock or scavenger crabs are out to grab with their sharp claws any bit of edible stuff they can handle that has been brought in by the sea. Sometimes hundreds of them make such a clattering and show so many dark forms scuttling sideways in various directions that much of the beach seems crawling with them! More about noises on the beach will be found in Chapter 9.

What to do. Just stay still for a bit at any spot on the beach and life will soon become active nearby, as only motion or the sound of footfalls bothers the small creatures. When there is a good low tide, it is possible to go down towards the sea through layer after layer of different forms of life, all merrily about their twice-daily work of getting food brought in by the tides.

CHAPTER TWO

Pinpointing Nature's Nocturnal Creatures

IN MOST AREAS in North America, but especially in the deserts, mammals are far more likely to be seen at night than in the daytime. Since their fur gives them good insulation against the cold that becomes greater after dark, while, on the other hand, the same fur or hair may make them uncomfortable during the heat of a summer day, many of them prefer the cooler hours of darkness. Another factor of great importance to mammals is the ease of hiding they find at night, so escaping many enemies and particularly man, who hunts mainly by day. The mammals that are pictured, described, and mapped in the Guide to Sights of Mammals at the end of this chapter are the ones most likely to be seen at night. Occasionally other mammals that like mainly daylight living, such as most of the tree squirrels and ground squirrels, will come out to play or hunt food on full-moon nights, but these are rare occurrences. Where such creatures as the yellow-bellied marmot show some frequency of night activity, they are listed in this book. Often such increasing frequency of using the night hours is caused by hunters. Daylight animals, finding themselves too strongly hunted, find night the best escape.

Your attitude while walking or sitting in the outdoors may well help or hinder the chances of seeing creatures, particularly the mammals, who are extremely sensitive to smells or sights that may convey emotions to them. If one is afraid or nervous, many animals can smell this, or can see nervous motions, which cause them in turn to become upset and to move quickly away or hide. Become instead calm and collected, moving, when necessary, very deliberately and without any hasty motions. Naturalists, who have coated their bodies and clothes with the smell of a strong-scented wild herb, or who have even covered themselves with the smell of the animals that they want to watch, have been able, by keeping cool and calm, and moving slowly and in harmony with their surroundings, to gradually approach and finally merge with certain herd animals, such as deer, and approach very closely many other creatures without alarming them.

Being in the Right Place at the Right Time

Mammals are creatures of habit and are often selective as to the times of night when they are most active. Thus certain bats, such as the pipistrelle and the pallid bat, come out with the first dusk and tend to vanish later, while other bats, such as the big brown bat and the silver-haired bat, like to hunt insects later in the night. Most gophers seem to be most active in the early evening or just before dawn, and will come out of the ground at such times to get a mouthful of green plants and then duck back quickly into their holes to eat it, or if recklessly brave from hunger, stay outside to eat it and then quickly grab another mouthful. Knowing that this is the time to see them, place a chair near their holes and become very quiet and calm. Remember also that other creatures, such as wildcats, foxes, owls, and so forth also know this characteristic of gophers and may come looking for them, so providing interesting sights to watch as the hunters creep up on and try to catch the hunted. Knowing the habits of the mammals of the night can thus often mean the difference between seeing them in action and missing them entirely by coming at the wrong time or to the wrong place.

Droppings and Tracks: Clues to Animal Whereabouts

Mammals leave their droppings (dung) and their tracks in many places to tell much about their habits. Finding in the daytime fresh droppings and tracks of animals that are out mainly at night tells where to watch when dusk brings its long shadows. A mouse, rat, wildcat, or fox may and often does become a creature of habit about where it leaves its droppings, choosing over and over again a favorite place, particularly a place where it is hidden, where it can watch for enemies or prey, and where there are one or more good escape routes. Finding such a pile of droppings gives the chance to zero in on it, by placing a campchair or piece of foam rubber in a place not too near or not too far away to watch at night, particularly with field glasses and using a red flashlight, and see the animal when it comes to its dropping place. But be sure to be quiet.

How to Identify Mammals Systematically

Identifying mammals at night is obviously hard because of the darkness. The silhouettes shown with the descriptions later in this chapter give an idea of shape and so what to look for. Note particularly length of tail and ears and shape of head and body. Thus a fat ratlike animal with comparatively small ears and practically no tail is obviously either a pika (related to the rabbits) or mountain beaver, as these are the only two animals that fit this description. To divide them further, notice color and habitat. The pika usually has rather light-colored fur because it gathers food in open meadows and around usually light-colored rocks, and is very rarely out at night, while the mountain beaver has dark fur because it is generally found in the thick growth near streams in deep, dark forests and is fond of darkness. So each animal can be identified instantly by color, habitat, and habit.

Mice and rats are particularly hard to identify because so many look alike. The Guide to Sights of Mammals refers to many of them by groups or genera rather than by species because, in most cases, to try to help identify species in the nighttime would be futile. However, some species do have special qualities that stand out and these are emphasized when useful, as, for example, with the bushy-tailed wood rat, which is the only kind that shows this kind of tail. Watch especially for lengths of tails and kinds of hair on tails. Many species can best be identified by habitat or by locality (see maps). Thus, among the mice, *Phenacomys* and *Pitymys,* or tree mice, live mainly in the woods, *Microtus,* meadow mice, in meadows.

It is best to get the feel of each animal habitat you watch by observing how the mammals adapt to it. Thus porcupines are found near trees, especially pines, up which they climb to be safe from enemies and to strip the bark. Both tree mice and tree squirrels build nests in the trees, the former smaller nests, but occupied by day instead of by night, when the mice go foraging. In a meadow, the maze of field mice or vole tunnels through the grass and under the ground forms highways for insects, birds, and predators, like snakes and weasels, that come after the mice. Note the totally different ways different predators catch the mice; the fox generally by pouncing, the wildcat by patiently waiting, the weasel by tracking.

So the puzzles and truths of life unfold in the darkness, where mammal life, at least, is moving about in fascinating numbers on innumerable adventures—tragedies and triumphs of escape or seizure.

Key to Using the Guide to Sights of Mammals

It is important to know what may be seen in the immediate surroundings wherever you are. In the following list of common mammals active at night, the species are placed in their natural groups. In many cases species are listed, but often just the genus (a grouping of several similar species) is given where the members of that genus are hard to distinguish except by experts; and sometimes several similar genera are lumped together.

Often similar species within a genus may be so alike that it would be foolish to try to differentiate between them here, but they are separated from other species in the same genus that are sufficiently different to be told apart from them. Thus all cottontail rabbits of the genus *Sylvilagus* are put together under the common name "cottontail," but the brush rabbit, which also is a part of the *Sylvilagus* genus, is mentioned separately because it looks quite different from the cottontails.

As each species or genus (when given as a whole) or part of a genus (that all look alike) is mentioned, the most important characteristics differentiating each species or group are described; then their habitats are given. Next, the most likely sights at night are mentioned. There are also maps showing in what parts of North America the creatures described are found.

The user should go through this list with a colored pen and mark with a checkmark or underline the names of all species and genera that are found in his part of the country as shown by the maps, and underline with the same colored pen all habitats found in his vicinity that are mentioned with these local species and genera. This makes it possible for him to look up much more quickly those creatures and plants in his own neighborhood.

Just keep thumbing through these pages to gain familiarity with all the kinds of night creatures described and mapped for your locality. Books listed in the Suggested References will supply more detailed knowledge. Then be on the watch for nature's fascinating nocturnal creatures when darkness comes.

GUIDE TO SIGHTS OF MAMMALS*

Order of Marsupials, with pouches on body for carrying young, and long prehensile scaly tails.

Opossum

Common Opossum *(Didelphis marsupialis)*. Gray body, white nose, clawless thumb, long naked scaly tail. Length of body: 15-21″. Habitats: farming areas, towns, open woods, streamside woods, generally at low elevations. Sights: seen climbing in trees, robbing garbage pails, feigning death when attacked (called "playing possum"), hanging by tail from branch.

Order of Insectivores. Primitive mammals with long noses, all teeth similar and sharp. Generally small size (all under 1′ long). Smooth velvety fur; have large powerful front feet for digging (except shrew mole). Moles and shrew moles.

Shrew Mole

Mole family, Talpidae.

Shrew Mole *(Neurotrichus gibbsii)*. Front feet longer than broad; velvety black hair with silver hairs on top of back. Habitats: coniferous forests of Northwest, generally near streams. Sights: digging subsurface runways in loose soil and leaf mold; rarely seen on surface.

Moles with front feet broader than long.

Western Mole

Western Moles *(Scapanus species)*. Length about 6-7″. Naked tails and very dark fur. Habitats: fields, meadows, lawns, open places in woods, streamside woodlands, where soil is fairly damp and loose. Sights: usually show only clod-filled mounds and ridges of loose soil pushed up as mole swims through dirt (seen doing this mainly at night). Can be captured by shoveling under at point of movement.

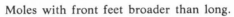

* Lengths given are of body and head only, unless otherwise specified. Tail length may be given separately.

Star-nosed Mole

Range of Opossum

Ranges of Western and Hairy-tailed Moles

Ranges of Shrew and Star-nosed Moles

Star-nosed Mole *(Condylura cristata)*. Fleshy, fingerlike tentacles around nose. Length: 5-6½". Habitats: wet ground near ponds, lakes, streams. Sights: digs along just under surface at night; may come out for brief swim in water.

Eastern Mole

Eastern Mole *(Scalopus aquaticus)*. Light brown or whitish colors, fur silvery; tail naked; nostrils open upward. Length: 4½-6½". Habitats: sandy soil of gardens, meadows, streamside woods. Sights: soil moves as mole swims in it just under surface. May swim in water.

Hairy-tailed Mole

Hairy-tailed Mole *(Parascalops breweri)*. Identified by heavily-furred tail. Length about 4-4½". Habitats: mainly mountainous forests in clearings; rarely in lawns or gardens of nearby towns. Sights: more likely than other moles to come out at night and wander over ground. Earth seen moving as it swims through subsurface.

Least Shrew

Shrews, family Soricidae. Have small feet. Size usually quite small, under 4". Always told from mice by sharp-pointed nose and the fact that the ears are concealed in the fur of the head.

Short-tailed Shrew

Shrews with tails shorter than 1¼" and less than half the length of head and body.

Least Shrews *(Cryptotis* species). Very tiny, 2-2½"; reddish brown color. Habitats: open grasslands. Sights: running in and out of prairie mole holes.

Northern Water Shrew

Short-tailed Shrew *(Blarina brevicauda)*. Eyes almost invisible. Length: 3-4"; purplish brown color. Habitats: almost everywhere where there is some dampness, but prefers forests and low swampy places. Sights: comes out of subsurface burrows for short trips at night after insects.

Shrews with tails longer than 1¼" and longer than half the length of body and head.

Large shrews of the genus *Sorex,* generally more than 3½".

Pacific Water Shrew

Northern Water Shrew *(Sorex palustris)*. Heavily fringed with hair along hind foot; body dark gray above, silvery below. Habitats: small streams and ponds. Sights: seen swimming under water with trail of silvery bubbles, or skittering across surface.

Pacific Water Shrew *(Sorex bendirei)*. Hind feet a little less fringed than above shrew, but color dark brown or black all over. Habitats: wet places and streams. Sights: swimming at night, or skittering on surface of water.

**Range of
Northern Water Shrew**

Range of Eastern Mole

Range of Least Shrew

**Range of
Short-tailed Shrew**

**Range of
Pacific Water Shrew**

Small Shrew of Genus Sorex

Small shrews of the genus *Sorex* and *Microsorex*, generally less than 3⅓" long, brownish colors. Since these shrews are hard to tell apart, they are lumped together here. Most prefer damp habitats in or near woods and brush. Sights: seen hunting insects, worms, mice, etc. at night quite voraciously and ferociously.

Order of Chiroptera. Bats (only mammals with true flight, having membranes between legs, and legs and tail). Bats are different from birds in that they catch insects one by one by tracking them down with high-frequency sound echoes bouncing off the insect prey!

California Leaf-nosed Bat

Leaf-nosed bats with leaflike projection on nose and third finger of wing with 3 branches.

Bats of family Phyllostomidae. Example: California leaf-nosed bat *(Macrotis californicus)*. Leaflike on nose; length with tail up to 4", pale brownish gray color. Habitat: caves and desert areas of Southwest. Sights: capturing insects in air at dusk.

Bats without leaflike projections on nose and third finger of wing with only 2 branches. Tail almost completely attached to wing membranes. Common bats of family Vespertillionidae.

Pallid Bat

Bats with high narrow ears.

Pallid Bat *(Antrozous pallidus)*. Size with tail: 4-5"; color light gray to yellowish brown. Habitats: generally low open country, deserts, plains, grasslands, brush. Sleeps by day in caves and buildings. Sights: catching flying insects in dusk.

Spotted Bat *(Euderma maculatum)*. Ears curved near tip, 3 whitish blotches on the blackish back. Size with tail about 4". Habitats: mainly arid open country. Rare.

Big-eared Bat *(Plecotus townsendii)*. Ears very long, straight. Size with tail: 3½-4"; brownish colored, lump on nose. Habitats: open woods, grasslands, sagebrush-deserts, etc. Sights: catching flying insects at dusk.

Range of Small Shrews of Genus Sorex and Microsorex

Long-eared Myotis *(Myotis evotis)*. Small size, light golden brown color. Habitats: mainly in woods.

Range of California Leaf-nosed Bat

Spotted Bat

Big-eared Bat

Long-eared Myotis

Range of Pallid Bat

Range of Spotted Bat

Ranges of Western and Eastern Big-eared Bats

Range of Long-eared Myotis

Bats with more rounded ears.

Little Brown Bats *(Myotis species)*. Small size, less than 4" total length. Most of these are brownish in color. Habitats: greatly varied. Sights: commonest bats seen at night, often flying over water.

Pipistrelle *(Pipistrellus hesperus)*. Tiny, 2½-3" total length, light yellow gray color. Habitat: deserts. Sights: drinking at first dusk.

Bats of larger size, more than 4" total length.

Hoary Bat *(Lasiurus cinereus)*. 5-6" total length, brown to yellowish brown, with white-tipped hair on top of back; narrow wings and large size distinctive. Habitats: generally near woods, meadows, lakes. Sights: very erratic flight.

Silver-haired Bat *(Lasionycteris noctivagans)*. About 4" total length; distinctive blackish color, touched all over with silvery white. Habitats: mostly in forested country. Sights: seen hunting insects over water.

Red Bat *(Lasionycteris borealis)*. About 4" total length, bright reddish or rusty red color, but often frosted with white-tipped hairs. Habitats: wooded areas. Sights: roosting in trees.

Big Brown Bat *(Eptesicus fuscus)*. 4-4⅓" total length; pale brown to reddish brown color, nose and ears blackish. Habitats: open woods and fields, caves and buildings. Sights: seen flying later at night than other bats, catching insects in open areas.

Evening Bat *(Nycticeius humeralis)*. About 4" total length; dull brown color; thick, leathery, small black ears. Habitats: woods. Sights: very swift flight in early evening.

Little Brown Bat

Pipistrelle

Hoary Bat

Big Brown Bat

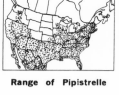

Silver-haired Bat   **Red Bat**

Evening Bat

Range of Little Brown Bat

Range of Pipistrelle

Range of Silver-haired Bat

Range of Red Bat

Range of Big Brown Bat

Range of Evening Bat

Range of Hoary Bat

Order of Lagomorpha. Hares, rabbits, pikas. Usually long legs and long ears. 8 large teeth for gnawing in front, instead of 4 as in rodents. Length usually more than 11".

Snowshoe Hare

White-tailed Jackrabbit

Black-tailed Hare

Antelope Jackrabbit

Cottontail Rabbit

Snowshoe Hare or Varying Hare *(Lepus americanus)*. Total length: 16-18"; length of ear: 3½-4"; grayish brown above and whitish below in summer; usually white except for black-tipped ears in winter. Habitats: coniferous forests, meadows. Sights: leaping along trail through woods.

White-tailed Jackrabbit or Hare *(Lepus townsendii)*. Total length: 22-24"; length of ear: 5-6"; tail whitish except for grayish stripe on top; color grayish in summer, whitish all over in winter. Habitats: grasslands and sagebrush areas. Sights: may hop about on moonlit nights; runs across roads at night.

Black-tailed Hare or Jackrabbit *(Lepus californicus)*. Length: 17-21"; length of ear: 6-7"; grayish brown in color, with a black streak on tail above, and large black-tipped ears. Habitats: grasslands, open brushlands and woodlands, sagebrush plains, etc. Sights: running across roads at night, hopping through brush in moonlight.

Antelope Jackrabbit *(Lepus alleni)*. Length: 19-22"; length of ear: 7-8". The large ears without black on top and the whitish sides and hips are distinctive. Habitats: mesquite-covered hills and plains, cactus deserts, brushy hills. Sights: noted for its enormous leaps, making it appear like a white flash in moonlight.

Cottontail Rabbits *(Sylvilagus species in part)*. Length: 11-17"; length of ear: 2-4". White cottony tail is distinctive. Back usually is dark grayish marked with reddish brown above. Habitats: eastern cottontail, near or in heavy brush or woods, often in briar patches; mountain cottontail, near or in brushy areas of western mountains; Audubon or desert cottontail, desert scrub and thickets on edges of woodlands, grasslands, canyon bottoms;

Range of Snowshoe Hare

Range of Black-tailed Hare

Ranges of White-tailed and Antelope Jackrabbits

Ranges of Eastern Cottontail and Brush Rabbits

Ranges of Mountain and New England Cottontail Rabbits

Range of Audubon Cottontail Rabbit

New England cottontail, brush and woodland areas of mountains. Sights: hopping about and dancing or fighting with high kicking in clearings on moonlit nights.

Marsh and Swamp Rabbits *(Sylvilagus species in part)*. Length: 14-17"; length of ear: 2-3"; brownish rabbits with tail indistinct. Habitats: swamps, marshes, and wet areas near streams. Sights: splashing through shallow water areas, or swimming in moonlight.

Marsh Rabbit

Brush Rabbit *(Sylvilagus bachmani)*. Length: 11-13"; length of ear: 2-2½"; a small brown rabbit of Far West. Habitat: distinctively brushy chaparral. Sights: feeding on edges of brush in the moonlight.

Brush Rabbit

Order of Rodentia. Rodents (noted for 4 large front gnawing teeth or incisors). Squirrels, mice, etc.

Five rodents uniquely different from all the others.

Mountain Beaver *(Aplodontia rufa)*. Tail so small it is all but invisible. Length: 14-15"; chunky, round body; dark brown except for white spot below ear; powerful legs for digging. Habitat: humid coastal forests, usually in rocky thickets of wild berries, digging tunnels just below surface. Sights: comes out at night to eat plants near holes, even climbs fir trees to eat needles, and swims in pools or streams.

Mountain Beaver

American Porcupine *(Erethizon dorsatum)*. Length: 18-23"; length of tail: 7-9"; yellowish brown to blackish above; body and tail covered with sharp quills. Habitats: coniferous forests, pinyon-juniper woodlands. Sights: up in trees eating bark; around camps or cabins chewing on salt-impregnated axe handles, etc.; bristling up sharp quills and thrashing tail to resist attack.

American Porcupine

Nutria *(Myocastor coypus)**. Length: 22-26"; looks like enormous ugly brownish yellow or reddish brown rat, with round scaly tail; hind feet very large with webbed toes and long sharp claws. Habitat: large streams, streambanks and nearby fields, marshes. Sights: swimming in water, using tail to splash water as signal of warning.

Nutria

Beaver *(Castor Canadensis)*. Length: 24-30"; tail: 9-11". Large scaly tail flattened from top to bottom, not sideways; dark brown

Beaver

---

* Information about the nutria is not sufficient to furnish a map of its geographical distribution.

Ranges of Marsh and Swamp Rabbits

Range of Mountain Beaver

Range of American Porcupine

Muskrat

Flying Squirrel

Yellow-bellied Marmot

Woodchuck

Pocket Gopher

to golden color; large webbed hind feet. Habitat: in or near streams and ponds with willows and/or aspens and cottonwoods. Sights: V-shaped ripple on pond with head showing at tip of V; loud slapping of tail on water as warning; work on canals and on dam; gnawing down of trees for food or for use in dam by cutting out large chips with teeth.

Muskrat *(Ondatra zibethica)*. Length: 10-15"; tail: 8-11"; distinctive tail flattened from side to side for swimming; generally brown color; large webbed hind feet. Habitat: streams, ponds, marshes, nearby fields and woods; makes holes in banks, and brush huts on water among cattail rushes. Sights: V-shaped ripple on water with dark head at night; swishing of tail from side to side; working at building nest house.

Common types of rodents.

Squirrels, family Sciuridae. Marmots, tree squirrels, ground squirrels, etc.

Flying Squirrel *(Glaucomys* species). Length: 5½-6"; tail: 3½-5½"; glossy olive brown fur; skin stretches between legs to allow animal to glide (not fly). Habitat: dense forests. Sights: beautiful when seen gliding from tree to tree in moonlight. Our only truly nocturnal squirrel.

Yellow-bellied Marmot *(Marmota flaviventris)*. Length: 14-20"; tail: 4½-9"; thick heavy body is yellowish brown with grizzled hairs and yellow belly; dark mark across whitish face. Habitat: rocky slopes near alpine meadows, also in lower grasslands where there are rocks. Partially nocturnal near towns. Sights: feeding on plants near rock nest, making dash for shelter when alarmed.

Woodchuck *(Marmota monax)*. Length: 16-20"; tail: 4-7"; reddish brown to blackish color; no bar on nose; black feet; chunky body. Sometimes comes out on moonlit nights or becomes nocturnal when overhunted. Sights: feeding on plants near hole, or dashing for hole when alarmed; vicious fights between males.

Pocket gophers, family Geomydae. Large front digging legs; fur-lined pockets on outside of cheeks to carry food; fine dirt of digging pushed out into rough pile, not conical like mole's.

Pocket Gophers *(Thomomys* species, *Geomys* species, and *Cratogeomys castanops)*. Length: 4-9"; tail varies; general brown color but varies from almost white to black; very large front teeth.

Range of Beaver

Range of Muskrat

Range of Flying Squirrel

Range of Yellow-bellied Marmot

Range of Woodchuck

Range of Western Pocket Gophers (Thomomys species)

Gophers are hard to tell apart; use maps and other books. Habitat: open areas. Gophers prefer grasslands, but range from deserts to open places in woods. Sights: pushing dirt up out of burrows into distinctively shaped mounds; pulling plants into burrows; traveling short distances on surface to get plant foods or find mates.

Spiny Pocket Mouse

Pocket mice, kangaroo rats, and kangaroo mice, family Heteromyidae. Small to large mice with long tails, small ears, gray to brown colors above and white below, and fur-lined cheek pouches. The front feet are rather weak, but the hind feet are very strong, adapted for jumping; fur rough to feel, usually with bristly spines on rump.

Pocket mice. Hind legs and tail not especially long for big jumps though larger than in most other mice.

Spiny Pocket Mouse *(Perognathus spinatus)*. Length: 3-3⅘"; tail: 3½-4½"; brown and white spines on rump; body grayish brown above, sometimes black. Habitat: desert scrub with rocks. Sights: seen jumping about among rocks and bushes.

Desert Pocket Mouse

Desert Pocket Mouse *(Perognathus penicillatus)*. Length: 3-3¾"; tail: 3½-3¾"; dark crest on end of tail; yellowish brown to gray in color above. Habitat: low desert scrub. Sights: takes large leaps in moonlight.

Hispid Pocket Mouse *(Perognathus hispidus)*. Length: 4½-5"; tail: 3½-4½"; tail shorter than head and body and without crest on end; grizzled brown color. Habitat: grasslands and desert scrub. Sights: jumping about in moonlight.

Hispid Pocket Mouse

Smooth-furred Pocket Mice *(Perognathus* species in part). Length: 2-5"; tail: 1½-5". Habitat: mainly dry open areas of grasslands and deserts, but sometimes found in brushlands or chaparral. Sights: fast or jumping movements at night, seeking food or escaping enemies.

Mice with inner cheek pouches, but tails and legs unusually large.

Kangaroo Rats *(Dipodomys* species). Length: 4-6"; tails: 5-8½"; very attractive-looking creatures, not deserving the name of "rats," with very long and powerful hind legs, small weak front legs, and usually very long tufted tails. Habitat: usually dry open areas, though some like dense brush. Sights: make tremendous jumps, some fighting in midair by powerful kicks;

Smooth-furred Pocket Mouse

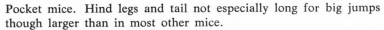

Range of Eastern Pocket Gophers (Geomys species)

Range of Mexican Pocket Gopher (Cratogeomys castanops)

Range of Desert Pocket Mouse

Kangaroo Rat

Ranges of Spiny and Hispid Pocket Mice

Range of Smooth-furred Pocket Mice

Range of Kangaroo Rats

Dark Kangaroo Mouse

Pallid Kangaroo Mouse

Harvest Mouse

White-footed Mouse

Grasshopper Mouse

**Range of
Dark Kangaroo Mouse**

**Range of
Pallid Kangaroo Mouse**

dive into holes at high speed when danger approaches; hunt seeds, from which they get both moisture and food.

Dark Kangaroo Mouse *(Microdipodops megacephalus)*. Length: 2½-3″; tail: 2¾-4″; similar to kangaroo rat in appearance but much smaller and with tail thick in middle; dark brown or black in color above, to look like dark sand. Habitat: dark sandy desert lava soils. Sights: takes big jumps in moonlight; hunts seeds.

Pallid Kangaroo Mouse *(Microdipodops pallidus)*. Length: 2½-3″; tail: 3-4″; light reddish-brown pink above, white below; otherwise as above. Habitats: light sandy desert lava soils. Sights: as above.

Mice and rats without pockets in their cheeks, family Cricetidae and family Muridae. Generally do not make big jumps as the above do.

Harvest Mice *(Reithrodontomys* species). These are the only cricetine mice with grooves in their two large upper incisors or forward teeth; otherwise they look like house mice except for a generally cleaner, more attractive appearance. Length: 2-3″; tail: 1⅝-2⅗″; brownish to blackish above, grayish and often tinged with yellowish brown below. Habitat: fields, meadows, woods' edges. Sights: running about at night in many areas, usually in low dense plant growth, feeding on seeds and fruits.

White-footed Mice *(Peromyscus* species). These medium-size mice (larger than house mice) have white feet and usually white bellies, being brown or yellowish brown above. Length: 3-5″; tails: 2-5″; tail always covered with usually short hair (whereas house mouse has naked tail); generally much more attractive-looking than house mouse. Habitat: almost everywhere, often nesting in holes in ground, under rocks and in abandoned buildings. Sights: running about with bright, nervous actions; feeding mainly on seeds and insect grubs.

Grasshopper Mice *(Onychomys* species). Easily told by short white-tipped tails, chunky appearance and pale colors. Length: 3½-5″; grayish to pinkish above, white below. Habitat: sagebrush or valley grasslands in West. Sights: hunting large insects and even lizards and small mice, attacking viciously; following trail like wolf.

Rice Rat *(Onyzomys palustris)*. Length: 4½-5″; long scaly tail 4½-7″; rather short round ears; short fur. Habitat: moist grassy and sedgy areas near water. Sights: running to hide, or to hunt seeds and insects.

**Range of
Harvest Mice**

**Range of
White-footed Mice**

**Range of
Grasshopper Mice**

Cotton Rats *(Sigmodon* species). Distinguished by strongly yellowish-brown grizzled gray-brown to black-brown fur. Length: 5-8"; tail: 3-5". Dark above, pale below. Habitat: high, damp grass, filled with their runways. Sights: running along runways and feeding on plants.

Rice Rat

Wood Rats *(Neotoma* species). Length: 6-9½"; tail: 4-9½"; easily distinguished from similar-sized Norway rat by hairy rather than scaly tails, large ears, white feet and belly, and by habit of making stick, rubbish, or cactus parts nests. Habitat: woods, brush, desert scrub. Sights: seen hauling bright objects or seeds into large nests; often called trade rats because of trading one object for another. (Bushy-tailed wood rat, *Neotoma cinerea,* is distinguished from all others by large bushy tail.)

Cotton Rat

Bog Lemmings *(Synaptomys* species). Length: 3-7"; tail usually shorter than body; grayish brown to brown in color; ears very small, sometimes invisible. Habitat: usually thick grassy vegetation, meadows, and grasslands of moist areas. Sights: running along runways in grass or feeding on grass.

Sagebrush Vole *(Lagurus curtatus).* Length: 3½-4½"; tail: ½-1"; distinctive ashy gray color above, white below. Habitat: sagebrush. Sights: running along runways in sagebrush, or diving into its hole.

Wood Rat

Pine Vole *(Pitymys pinetorum).* Length: 3-4"; tail: ¾-1"; distinctive thick soft reddish brown fur. Habitat: floor of eastern broad-leaf forests, some pine forests. Sights: running in and under fallen leaves and leaf mold.

Red-backed Voles *(Clethrionomys* species). Length: 3½-4¾"; tail 1½-2"; small grayish brown forest mice with reddish strip

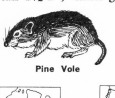

Pine Vole

Red-backed Vole

Bushy-tailed Wood Rat

**Range of
Rice Rat**

**Range of
Cotton Rats**

**Range of
Wood Rats**

Bog Lemming

**Range of
Bog Lemmings**

**Range of
Sagebrush Vole**

**Range of
Pine Vole**

Sagebrush Vole

Tree Phenacomys

**Range of
Red-backed Voles**

**Range of
Tree Phenacomys**

**Range of
Meadow Mice**

Meadow Mouse

Florida Water Rat

Jumping Mouse

House Mouse

down back. Habitat: usually on floors of the forests, but Gapper red-backed mouse is found in meadows, brush and rock slides.

Tree Phenacomys *(Phenacomys longicaudus)*. Length: 3½-4½"; tail: 1-3"; generally reddish brown above. Sights: build nests high in trees and run about on branches.

Meadow Mice *(Phenacomys* species in part and *Microtus* species). Length: 3-5¾"; tail: 1½-2½". Living in grassy areas, meadow mice generally have soft, grizzled long grayish brown fur and blunt noses. Habitat: grassy forest clearings or meadows. Sights: running in trails through grass, or diving down holes.

Florida Water Rat *(Neofiber alleni)*. Length: 7-8½"; tail: 8-11"; tail black and flattened side to side; easily told by dense velvety brown fur above, and silvery fur on belly. Habitat: mainly in water or near water. Sights: swimming with trail of silvery bubbles near their conical reed and leaf houses built above water in the marshes.

Jumping Mice *(Zapus)*. Length: 3½"; tail: about 5". Extremely long tail; long hind feet, weak front feet. Habitat: damp thick grassland. Sights: make great leaps through thick grass.

Old World rats and mice, with coarser gray-brown fur than above, and long, scaly, naked tails. Appearance of rats in particular is much uglier than our native rats and mice.

House Mouse *(Mus musculus)*. Length: 3-3½"; tail: 3-4"; tail as long or longer than body and head together; grayish brown above, grayish to yellow-brown below; naked scaly tail is uni-colored gray. Habitat: near or in buildings. Sights: running in cupboards or storage bins.

Norway Rat *(Rattus norvegicus)*. Length: 7-10"; naked tail 5-7½" shorter than head and body; overall grayish brown color. Habitat: cities, farms, buildings. Sights: running into drain holes, sewers, storage areas, etc.

**Range of
Florida Water Rat**

**Range of Western
Jumping Mouse**

**Range of Woodland
Jumping Mouse**

Norway Rat

Common Range of House Mouse and Norway Rat

Range of Black Rat

Black Rat

Black Rat *(Rattus rattus)*. Length: 6-7½"; naked tail 8-9½" always longer than head and body; may be brown or black, with grayish belly. Habitat: near or in buildings. Sights: usually found higher in buildings and seen running along wires and cables.

Order Perissodactyla. Animals with singled-toed hoof. Wild horses and burros.

Wild Burro *(Equus asinus)* *. 4' high at shoulder; erect dark mane on neck; long ears. Habitat: desert scrub, desert mountain meadows. Sights: often in herds feeding at night, or running through brush.

Wild Horse or Mustang *(Equus equus)* *. 5' high or more at shoulder; loose mane on neck; shorter ears. Habitat: grasslands, meadows. Sights: often in herds feeding at night.

Wild Burro

Order Artiodactyla. Animals with 2-4-toed hoof. Deer, elk, pig, moose, bighorn sheep, etc.

Piglike animals.

Wild Boar *(Sus scrofa)*. About 3' high; length about 5'; weight up to 600 pounds or more; grayish brown; bristly hairs on back; hairy ears; sharp tusks; naked nose; 4 toes on hind foot. Habitat: oak woodlands, chaparral in West, broad-leaved woodlands of East. Sights: rooting ground for roots and nuts at night; boar or sow with young may charge if disturbed.

Peccary *(Pecari angulatus)*. About 3' long; 2' high; smaller than above, and with 3 toes on hind foot. Habitat: hills covered with oaks, also brush. Sights: moving in small herds through cover at night; or rooting in ground.

Wild Horse

Wild Boar

* Information about this animal is not sufficient to furnish a map of its geographical distribution.

Range of Wild Boar

Range of Peccary

Peccary

Canadian Elk

Range of Canadian Elk

Range of Mule Deer

Range of
White-tailed Deer

Range of Moose

Range of American
Woodland Caribou

Mule Deer

White-tailed Deer

Moose

American Woodland
Caribou

Deerlike animals. Horns shed in late fall, grow anew in late spring with velvet cover.

Canadian Elk *(Cervus canadensis)*. Height: 4-5½'; males 700-1000 pounds, females 500-650 pounds; brown color, tawny rump; main beam of horn has unbranched tines. Habitat: open coniferous forests and meadows. Sights: feeding at night on grasses or tree and bush leaves; on moonlit night two males may lock horns in battle; males mark scent post or tree with special burr on horn.

Mule Deer *(Odocoileus hemionus)*. Height: 3-3½'; 4-6' long. Weight: 100-250 lbs.; color: grayish blue in winter; reddish brown in summer; tail black on top or black-tipped; ears large; antlers branch equally (not with prongs from main beam). Habitat: forest edges, brushy places, and meadows. Sights: feeding in clearings on twigs; jumping away explosively with 4 hoofs hitting ground all at once.

White-tailed Deer *(Odocoileus virginianus)*. 4½-5½' long; height: 3-3½'; bluish gray in winter, reddish brown in summer; tail like a large white flag when waved to signal danger; antlers with two main beams, each with separate prongs. Habitat: dense forest and brushy areas, also low marshy flood plains. Sights: runs instead of jumps, showing white flag of tail.

Moose *(Alces alces)*. Up to 10' long; height: 5-7'; snout overhangs mouth; pendant hairy "bell" on throat, large flattened horns. Habitat: dense forests, shallow lakes or ponds, marshy areas near forests. Sights: may be seen feeding in lake or pond, dipping head far under water and carrying water plants up to surface; or if frightened, running at ungainly gait through forest.

American Woodland Caribou *(Rangifer tarandus)*. Length up to 8'; height: 3½-4'; whitish to dark brown in color, with neck always white, also small white rump patch; feet large for walking on snow; antlers partly flattened but not as much as those of moose. Habitat: forest edges and muskegs. Sights: seen feeding on grass, sedges, and lichens that grow on trees, or moving in herds through forest.

Barren Ground Caribou *(Rangifer aecticus)*. Similar, but pale to almost white; generally found in open barrens, except in winter.

Barren Ground Caribou

Order of Carnivora. Carnivores, flesh-eaters with long sharp canine teeth.

Dog family, Canidae, including coyotes, wolves, foxes.

Red Fox *(Vulpes fulva)*. Length: 20-26"; tail: 14-17", very large and fluffy with white tip; black legs and feet; general color reddish or reddish brown, but sometimes black with frosted hair tips, or with black blotches. Habitat: partially open woods and brush; meadows. Sights: seen pouncing on mice or chasing rabbits in clearings.

Kit Fox *(Vulpes mecrotis)*. Length: 14-20"; tail: 9-12", tipped with black; ears very large; general color grayish or fulvous above, whitish below and inside ears. Habitat: sagebrush, desert scrub, Joshua tree woodland, alkali sink and valley grassland. Sights: seen flitting very rapidly around edge of light from camp-fire; or bouncing or slinking after rats and mice when hunting.

Gray Fox *(Urocyon cinereoargenteus)*. Length: 20-29"; tail: 10-16", with black mane and black tip; gray above, reddish brown below, white throat. Habitat: sagebrush, chaparral, brushy areas, dense broad-leaved woods. Sights: climbing tree, hunting mice and rabbits in small clearings, often by pouncing.

Coyote *(Canis latrans)*. Length: 32-38"; tail: 10-16", carried low when running; mainly grayish, but tinged here and there with reddish brown. Habitat: found most everywhere, but prefers open brushlands or woods, also meadows. Sights: running through brush; two chasing rabbit or hare but taking turns; 2-5 howling on hilltop.

Red Fox

Kit Fox

Gray Fox

Coyote

Range of Barren Ground Caribou

Range of Red Fox

Range of Kit Fox

Range of Gray Fox

Range of Coyote

Gray Wolf

Red Wolf

Mountain Lion

Canadian Lynx

Gray Wolf *(Canis lupus)*. Length: 3½-4½'; tail: 12-20", black-tipped and held high when running; broad head; dark gray to black in color. Habitat: wild forest areas and meadows. Sights: seen running in packs or alone on a trail, with tail held high.

Red Wolf *(Canis niger)*. Similar, but a little more reddish; range entirely different. Sights: slinking through brush.

Cat family. Retractable claws.

Mountain Lion *(Felis concolor)*. Length: 3½-4½'; tail: 29-36", dark-tipped; weight: 80-220 lbs.; tawny or grayish in color. Habitat: open woods, coniferous forests, chaparral and other brush; rocky areas, pinyon-juniper woodland, etc. Sights: seen sneaking through brush, or lying in wait for deer with tail tip gently moving; pouncing on deer.

Canadian Lynx *(Lynx canadensis)*. Length: 2½-3'; tail: 4", black-tipped; large tufts on ears; grayish color; large feet for snow walking. Habitat: coniferous forests. Sights: walking over snow, lying in wait for rabbits, rats, etc.; pouncing.

Bobcat *(Lynx rufus)*. Length: 2-2½'; tail: 5", barred black; short ear tufts, reddish brown color above, white below, speckled with black. Habitat: most habitats, but likes rocky areas in West, swamps in East. Sights: crouched motionless, watching for prey, or pouncing on mouse, rat, or rabbit; sometimes climbing trees.

Bear family and relatives. Hind feet put flat on ground; thick-bodied generally.

Black Bear *(Euarctos americanus)*. Length: 5-6'; very short tail; weight: 150-400 lbs.; color brown or black, but nose always tawny brown; thick body. Habitat: coniferous forests, meadows. Sights: getting into garbage cans; walking through woods; cubs climbing trees.

Grizzly Bear *(Ursus horribilis)*. Length 6-7½'; very short tail; weight: 300-900 lbs.; yellowish to dark brown in color with white-tipped hairs giving grizzly appearance; large hump appears at shoulders. Habitat: mainly in Rocky Mountains in coniferous

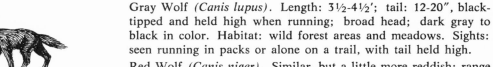

Ranges of Red and Gray Wolf

Bobcat

Black Bear

Grizzly Bear

Range of Mountain Lion

Range of Canadian Lynx

Range of Bobcat

Range of Black Bear

forests, meadows, and rocky areas. Sights: seen rolling over rocks to catch rodents; walking along trails. Should be avoided; stand still and watch. Flash light in face or climb tree if bear gets too near; above all, don't run.

Raccoon

Raccoon *(Procyon lotor)*. Length: 18-30"; tail: 8-12", with dark rings; grizzled grayish brown in color with black mask over eyes. Habitat: woods, usually near water. Sights: fishing in stream for frogs, fish, crayfish; washing food in handlike paws; lumbering along trail with rocking motion; climbing trees.

Ring-tailed Cat *(Bassariscus astutus)*. Length: 14-17"; tail: 14-16", fluffy and black-ringed; slender body grayish brown in color; ears large. Habitat: rocky, brushy areas. Sights: may be seen sneaking out from rocks to find food at night; moves very fast when frightened.

Ring-tailed Cat

Weasel family (Mustelidae). Rather slender or flat bodies and short legs.

Marten *(Martes americana)*. Length: 14-17"; fluffy tail: 7-9"; dense soft yellowish brown fur above, gray on face, yellowish orange on throat and dark brown behind and on tail. Habitat: coniferous forests and meadows. Sights: coming to meat bait on ground; climbing trees rapidly after squirrels.

Marten

Fisher *(Martes pennanti)*. Length: 20-26"; tail: 13-16", fluffy; general dark brown all over. Habitat: coniferous forests and meadows. Sights: rare, but one might be seen racing through trees at incredible speed after squirrels or martens.

Fisher

Mink *(Mustela vison)*. Length: 12-17"; tail: 5-9", round and thick but not fluffy; dark brown all over, but sometimes with whitish spots on belly and chin. Habitat: near or in water, especially marshes and fish-filled streams. Sights: swimming in water after fish or muskrat; eating fish on rock on shore.

Mink

Long-tailed Weasel *(Mustela frenata)*. Length: 8-11"; tail: 3-6", black-tipped; brown above, white below; very short legs and slender body. Habitat: most habitats, but likes rocks and brush.

Range of Grizzly Bear

Range of Raccoon

Range of Ring-tailed Cat

Long-tailed Weasel

Range of Marten

Range of Fisher

Range of Mink

Range of Long-tailed Weasel

Ermine

Range of Ermine

Range of Least Weasel

Range of Badger

Least Weasel

**Range of
Striped Skunk**

**Range of
Hooded Skunk**

**Range of
Hog-nosed Skunk**

Badger

Striped Skunk

Hooded Skunk

Hog-nosed Skunk

Sights: hunts mice and rats along small trails, also rabbits, bounding along in measuring worm-like leaps; dives into holes.

Ermine or Short-tailed Weasel *(Mustela erminea)*. Length: 5-9"; tail: 2-4"; similar to above weasel, but smaller and turns pure white in winter except for dark tip of tail. Habitat: generally high coniferous forests and meadows. Sights: chasing mice along runways or into holes. Pops in and out of holes at great speed.

Least Weasel *(Mustela rixosa)*. Length: 5-6½"; very short 1-1½" long tail is distinctive; colors as in weasels above, but no black tip on tail. Habitat: mainly coniferous forests.

Badger *(Taxidea taxus)*. Length: 17-22"; tail: 7-12"; flat body covered with rough yellowish gray fur; face marked with white and black; white stripe down back; powerful black front feet for digging. Habitat: desert scrub, sagebrush, chaparral, grasslands, woodlands, dry coniferous forests, meadows. Sights: waddling along low to ground; digging into ground to capture ground squirrels, etc.

Striped Skunk *(Mephitis mephitis)*. Length: 12-18"; tail: 7-11", fluffy and usually held high and curved at tip; black general color with white stripe (divided on lower back) on back and up forehead. Habitat: most habitats, but likes to be near streams. Sights: female leading kits down trail; stamping ground angrily and raising tail in warning; turning over rocks, sticks, etc. to find insects and grubs.

Hooded Skunk *(Mephitis macroura)*. Length: 12-16"; tail very long, 14-17"; general color black with faint white stripes on sides. Habitat: desert scrub and brush. Sights: as above.

Hog-nosed Skunk *(Conepatus leuconotus)*. Length: 14-20"; tail: 7-13"; hoglike snout for rooting; back and tail all white; face and lower parts black. Habitat: brushy areas, open woods. Sights: rooting for grubs in rotting wood, earth, and debris.

Range of River Otter

Steller's Sea Lion

River Otter *(Lutra canadensis)*. Length: 25-31"; tail: 12-18"; rich brown fur above, silvery beneath; feet webbed and tail thickened at base for use in swimming. Habitat: rivers, streams, marshes, lakes, and nearby woods. Sights: sliding down mud slides into water; rolling in grass to leave "sign"; diving and swimming rapidly in water with trail of silver bubbles.

River Otter

Family of eared seals and sea lions, Otaridae. Have small visible ears, while hind flippers can be brought up close to body.

Steller's Sea Lion *(Eumetopias jubata)*. Male up to 13', female up to 9' in length; large bull may weigh as much as a ton, the female less than half this. Yellowish brown color; young are dark brown. Habitat: rocky areas along seashore or in sea from San Francisco north. Sights: swimming in waves, or climbing out on rocky shore with rocking motion.

California Sea Lion

California Sea Lion *(Zalophus californianus)*. Male up to 8'; female to 6'; large bull may weigh ½ ton, female up to 600 lbs.; male dark brownish or blackish brown; female light brownish. Habitat: rocky areas along the Pacific coast. Sights: swimming in sea or climbing on rocks in moonlight; sometimes climbing out on sandy beaches at night.

Harbor Seal *(Phoca vitulina)*. Male about 6' long, female about 5'; weighing up to 325 lbs.; yellowish with black or grayish blotches, or grayish and spotted white. Habitat: Pacific, Atlantic waters, sometimes appearing in large rivers. Sights: swimming in herds near shore, or individuals crawling out on beach like giant caterpillars in undulatory motion.

Harbor Seal

CHAPTER THREE

Recognizing Birds of the Night in Action

BIRDS, UNLIKE THE mammals, are more commonly adapted to the daytime than to the night, but some—such as the owls, night herons, and whippoorwills—are especially built for nighttime hunting. Others—such as several of the shore birds, the nighthawks, and the chimney swift—seem equally at home by night or day, while still others—such as several thrushes, the ruffed grouse, and the martin—like the fringes of darkness in early evening or morning, but not the deep dark. Moonlight especially attracts the above birds plus others—like the various quail, the mockingbird, the cuckoo, and some of the wrens—to stay up much of a bright moonlit night singing, or food hunting. This may be especially true during the hot summer when such birds like to avoid the heat by retiring into deep shade during the day to sleep or rest.

The silhouettes, descriptions, habitats, and maps of bird distribution that are given in the Guide to Sights of Birds later in this chapter are to help identify birds seen in the dark. Shapes of bodies and tail and head are most easily seen, but watch also, when possible, for shape of bills and shape of feet. The comparatively medium size, thick body, and quick darting movements of plovers are distinctive, as compared to the more slender body, longer legs, and the ducking, bobbing, or wing-lifting movements of sandpipers and their relatives like the willet, the stilt, and the curlew. The forest habitats of most of the grouse (except the sage hen), plus their comparatively thick but long bodies make them distinctive.

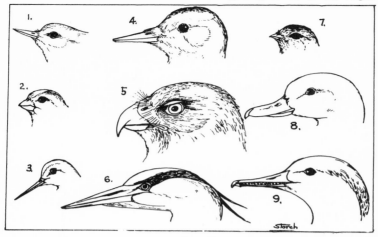

Types of Bills
1. Nuthatch   2. Finch   3. Shorebird   4. Woodpecker
5. Hawk   6. Heron   7. Swallow   8. Duck   9. Merganser

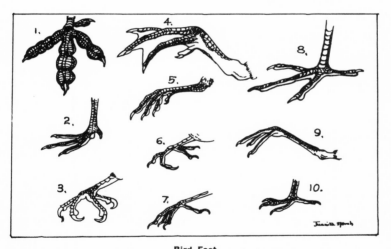

Bird Feet
1. Coot   2. Curlew   3. Woodpecker   4. Duck   5. Quail
6. Cuckoo   7. Sparrow   8. Heron   9. Killdeer   10. Pigeon

Telling Birds by Feeding Motions

Peculiar feeding motions and related actions are particularly useful in identification of some birds in the night. The avocet, for example, feeds by threshing the water with its bill, while the surface-feeding ducks dip their heads down and their tails up to feed on underwater plants while paddling rapidly. These distinctive ways of feeding, mentioned in the Guide to Sights of Birds at the end of this chapter, should be carefully watched for as they often immediately identify the bird seen.

Telling Birds by Geographical Location

Some birds, such as Vaux's swift and the chimney swift, look so much alike that you would be unable to separate them in the field, and especially at night, without knowing that the former is found almost entirely west of the Rocky Mountains, while the latter is found on the eastern side of the same mountains. Thus studying the maps to determine locations where such similar birds are found is often vital for correct identification.

Telling Birds by Habitat

Habitats become important in recognizing several species. Thus hardly more than a glimpse of a tiny owl in the southwestern deserts is needed to guess correctly that it is an elf owl. If the bird is seen popping in or out of an old woodpecker hole in a giant saguaro cactus, one can be even surer of its identity. Sagebrush is a sure and certain hiding place of the sage hen, a large partridgelike bird of the Great Basin deserts, while the blue grouse craves the shelter of the thick branches of a deep coniferous forest as contrasted to the ruffed grouse, who likes burnt-over or cut-over land with dead logs and thick bushes.

Telling Birds by Body Shape

In watching birds with binoculars at night, particularly note their body and head and tail shape against light areas—such as sky, open water, or open fields—that bring out the silhouettes of the dark bodies. Each usually has a distinctive shape, and there are few enough night-dwelling birds to make distinctions easy.

Nighttime Migrations

Something about flight patterns and times and ways of flight, especially when the birds are on migration, are given below. But remember that each usual way of night flight—such as hunting, playing, traveling, or courting—writes a separate sign against the night sky, a sign or distinctive diagram which anyone can soon learn to translate.

The great migrations of many birds from north to south, or south to north, are balanced by the smaller journeys of birds by night or dusk from place of feeding to home. Strangely, many day-flying birds choose the nighttime in which to migrate because of the lesser danger from predators. This is especially true of small birds—such as warblers, wrens, sparrows, vireos, orioles, flycatchers, and thrushes—who could easily be caught by hawks if they flew by daylight. Since they also fly high enough up in the sky when migrating at night to avoid the attacks of owls, which generally fly low when they hunt in the woods and fields, they find the night safer. No doubt the shore birds, and rails which also fly by night on migrations, do so for the same general reason, as many of them are slow fliers and easily caught by the falcons, which are the true tyrants of the bird world.

Another good reason for flying at night on migration is that the birds cannot catch food by night anyway because of the limitations of their eyesight in the dark; so it is better for them to fly by night and hunt what food they can by daylight hours. What about sleep? They probably do some sleeping by day in between bouts of food gathering. Most of the large water birds, such as ducks, migrate both by day and night, as many of them do know how to gather food at night.

Birds Silhouetted Against the Moon
1. **Great blue heron   2. Canada geese   3. Whippoorwill   4. Ducks
5. Hummingbird** (Note: Small birds often appear larger than big birds
when they are flying closer to the observer.)

Spotting Birds Against the Moon

Nighttime is a wonderful time to watch bird migration with a good pair of
binoculars or, better yet, a telescope that can magnify twenty times or up. The
telescope should be set on a tripod to steady it and taken out on full-moon or near
full-moon nights to watch the migrating birds flying across the face of the moon.

Most birds on migration rarely go above 3000 feet except when they have to
cross over high mountains. This makes it possible to spot quite a few of them
in a pair of binoculars crossing the moon and many more with a fair telescope.

Of course, many more birds can be seen migrating when the moon is low
in the sky than when it is high. This is because a low moon appears with much
more of the lower atmosphere, where birds fly caught in its surface light. So it
is always best to watch for birds against the moon when it is fairly low. Another
factor involved in watching birds and also bats crossing the face of the moon
is the nearness or accessibility of an area crossed by a well-used migration route.
Birds often seem to like certain definite areas to travel through. People living or
visiting close to one of these well-traveled routes are in luck for night birdwatching
during spring and fall migrations. Ask the ornithologists in the nearest university
or college biology department for information on local migration routes, or ask
them to show the routes to you on maps.

The Department of Biology of Louisiana State University offers information on joining an organization dedicated to scientific study of bird migrations by moon watching. It is a highly interesting scientific project, developing valuable information on the patterns of bird night migration and the numbers of different kinds of birds involved. Already it has been discovered that most birds that fly by night take a rest in the early hours of the night, then begin to rise into the sky in greater numbers around 10 P.M. and may reach their peak by midnight. There is more individual flying at night as compared to the great flocks often seen in daylight migrations. Look at the silhouettes of typical birds flying across the moon shown in this section. Use these to help identify what you see.

One of the queer things sometimes seen at night during migration is birds flying the wrong way: that is, south in spring, and north in winter. Obviously there is some reason for this, but scientists are still looking for more answers and perhaps you can help find them. Besides seeing the birds cross the moon, you can also often hear their cries as they pass overhead, as many fly quite low, though high enough to avoid trees and houses.

Flight Patterns in the Sky

Notice that certain birds make certain patterns in the sky. Some of these patterns are the ragged, fluttering flight of a sandpiper; the great V's of the ducks; the long line of pelicans, with each going up as the next one in line goes down, and so on.

Bats Compared to Birds as Migrators

Bats are actually much more successful migrators than birds, losing far fewer of their numbers during flight. This is evidently because they are both surer fliers, with their sonar system to keep them from hitting the wrong objects, and more experienced in night flying. They also seem to avoid storms better and may have a built-in warning system about changes in the atmosphere brought on by the approach of storms. Their success is also partly explained by the fact that their females each have only one baby bat a year, instead of the at least three or four baby birds hatched per mother. The baby bats, clinging to the fur of their mothers from birth, are far better protected from falling and injury and danger of predators than baby birds. To top it all, bats are more efficient insect catchers while in flight than even swifts and swallows, and can keep themselves supplied with such food even when traveling in migration. One of the most interesting things to watch is a bat drinking from a stream or pond while in full flight by simply opening his mouth and scraping a bit of water off the surface.

But the birds probably make up for these bat superiorities by making longer and more sustained flights. Even tiny warblers, each weighing only an ounce, fly north from South America across more than 500 miles of open sea, often through bad winds and storms, and continue on in a frenzy of flight until they reach the southern United States, then move northward in waves into New England and Canada. How such a tiny mite of flesh and feathers can maintain a sustained speed of flight for so long is one of the great miracles of nature and of avian determination!

The last spring storms have scarcely died away and the first leaves appeared on the trees and bushes, before the world of the north seems suddenly flooded with the beautiful voices and songs of the returning migrators, and the flashing of yellow, green, and blue wings. The caterpillars and other bugs are now busy eating the leaves; and the warblers, thrushes, orioles, and other spring migrants are right there on

Bird Flight Patterns
Top: **Sandpiper.** Center: **Flicker.** Bottom: **Pelicans.**

time to start eating them and save the trees. The magic of their coming is stunning when we see none of them one day and then the next they are there, having come in migration out of the night, like a rabbit pulled out of a hat by a magician.

Nonmigrating Night Flights

There are other flights of birds at night that are quite remarkable. Where does the male chimney swift go when he leaves his mate on her nest in late spring evenings to wing his way into the darkening sky and disappear the whole night through? Sometimes he has to drive her back to the nest so he can go alone, but go he does on his solitary flight into the nowhere of the high dark, like a man going on a night binge in defiance of a happy marriage. Watch for him in the dawn and you may see him come out of the sky with the first gray waves of light, his feathers a little bedraggled and a tired look about him of one ready for rest.

Some of the shore birds, like the killdeer and the common snipe, are extraordinary fliers in the early part of the night during spring courting. I have heard them crying and calling above me in the darkness and listened to the swish of their wings as they sweep low to the earth and then up again in a speed that hurtles them in wide sweeps and barrel turns through the gloom. Imagine the glorious fun of their flight when tied in with the mating urge: the chase, the swirl, the turn, the triumphant capture on the ground or high in the sky, each feather in the whistling wings singing a song of love.

Down by the marshes and the lakes evening time is a good time to watch for the black-crowned and yellow-crowned night herons as they gather in flocks from the nesting trees to fly to their fishing grounds in the estuaries, pools, and lakes. As their great wings flap, the air is filled with loud rustling and off they go in long, straggling lines or sometimes V's. The night becomes filled with their cries, and one is astonished to be told that when they go hunting for frogs and crabs, crayfish and fish in the shallow waters, they are among the most silent of birds, moving forward with the stalking skill of the mountain lion.

Owls hunting at night are often told by the way they move, even without seeing their shapes or colors. The great horned owl, fiercer and more dangerous than even the larger great gray owl, flies through the woods and open meadows at night with the dash and bravado of one who is king of all he surveys. Even though he comes on extraordinarily silent wings, the better to get close to his prey, every move of his flight expresses boldness and ferocity. The great gray is confident enough, but somehow more clumsy and hesitant, as if he realized that most of his size is due to fluffed out feathers. So he is a skulker in the deepest parts of the woods rather than a sky pirate of almost every habitat, as is the great horned. The screech owl, on the other hand, is a most cautious soul, particularly as he knows very well that if caught in the open he is fair prey for the great horned. The screech stays among the branches of the trees, where he can dodge around a stem from his enemy, and drops down upon a mouse or large insect with the secretive air of one who is catching food without his master's permission. The short-eared owl, a larger bird, is fond of open marsh and grassland, where, on moonlit nights, it appears like poetry in motion, sweeping low back and forth on its wide wings over the ground in wavering flight that scares up mice and rats and rabbits and sends them scurrying frightened and bewildered until the sudden dropping blow of plunging talons strikes and kills.

And so it goes; each bird flying on the night breeze writes its own special message and song of flight against the dark sky.

GUIDE TO SIGHTS OF BIRDS

Owls, with large heads and eyes looking forward out of facial discs instead of from sides as in other birds. This is to help night vision. Strong claws are used to grasp prey; hooked bill tears prey.

Owls with "horns" or "ears" made of feathers on their heads.

Screech Owl *(Otus asio)*. Length: 8-10"; the only small owl with horns; may be either gray or reddish brown in color. Habitat: mainly woods, woods' edges and meadows. Sights: gliding down rather abruptly from tree to attack mouse or rat.

Great Horned Owl *(Bubo virginianus)*. Length: 20-24"; the only very large owl with horns; brownish gray on back, grayish in front with white collar and white X on face. Habitat: woods, coniferous forests, brush, desert scrub, meadows, grasslands, etc. Sights: long silent sweep on motionless wings to suddenly strike rat or rabbit on ground.

Long-eared Owl *(Asio otus)*. Length: 13-16"; a medium-sized owl with ears or horns set close together; slender brownish gray body, reddish brown face with white X over bill. Habitat: coniferous forests. Sights: seen motionless on limb, or gliding silently through branches.

Short-eared Owl *(Asio flammeus)*. Length: 13-18"; ears or horns rather small, just above eyes; general color brown, whitish on belly with brown streaks. Habitat: marshes and open grasslands. Sights: flies with irregular, flopping flight, low to ground, harrying rabbit or rat until it can drop on fleeing animal and seize it.

Owls without "horns" or "ears."

Barn Owl *(Tyto alba pratincola)*. Length: 15-20"; light brown above, pure white below and on monkeylike face; unusually long white legs. Habitat: open areas of grassland, brush or woods

Screech Owl

Great Horned Owl

Long-eared Owl

Short-eared Owl

Range of Short-eared Owl

Range of Barn Owl

Barn Owl

Range of Screech Owl

Range of Great Horned Owl

Range of Long-eared Owl

Snowy Owl

Range of Snowy Owl

Range of
Burrowing Owl

Range of Barred Owl

Burrowing Owl

Barred Owl

Spotted Owl

edges. Sights: flies like ghostly great moth in swoop to ground to seize mouse or rat.

Snowy Owl *(Nictea scandiaca)*. Length: 20-25"; very large white owl, lightly marked with dusky bars. Habitat: open areas, beaches. Sights: perched on fenceposts, haystacks, dunes, or other high points to watch silently for prey.

Burrowing Owl *(Speotyto cunicularia)*. Length: 8½-9½"; distinctively long legs; brownish color. Habitat: prairies, grasslands, desert scrub, brush. Sights: often seen on top of pile of dirt near hole, fencepost, etc., from which it may flop up and over in air to attack large insect or mouse; nods or bounces about when upset.

Barred Owl *(Strix varia* subspecies*)*. Length: 18-22"; head puffy-looking and unusually large, with numerous brown streaks against white; breast streaked white and brown crosswise; streaked lengthwise on belly; eyes very large and liquid brown. Habitat: mainly broad-leaved woodlands. Sights: silent, swift flights between trees or to ground to capture prey.

Spotted Owl *(Strix occidentalis)*. Length: 17-20"; similar to above, but found west of Rockies. Habitat: coniferous forests and canyons. Sights: as above.

Great Gray Owl *(Strix nebulosa)*. Length: 24-34"; very large in appearance, but bulk mostly caused by feathers; general gray color, streaked darker gray; very large facial discs. Habitat: deep coniferous forests. Sights: slow, powerful wingbeats followed by long glide just before attacking mouse or rat on ground.

Saw-whet Owl *(Aegolius acadicus)*. Length: 7-8½"; brownish, with brown streaks down white breast and belly, except that young owls appear dark brown with blackish face, and sharply marked white "eyebrows." Habitat: woods, coniferous forests, streamside woodlands. Sights: tamely sits on branch until you approach closely; flies softly like moth to ground to attack mouse.

Great Gray Owl

Saw-whet Owl

Range of Spotted Owl

Range of
Great Gray Owl

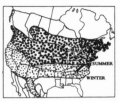

Range of Saw-whet Owl

Range of
Richardson's Owl

Range of Pygmy Owl

Richardson's or Boreal Owl *(Aegolius funereus)*. Length: 8½-12″. A large-headed small owl. Face disks framed with black; bill yellowish; forehead heavily dotted with white. Very tame. Habitats: Boreal forests of Northern Hemisphere in tree cavities, woodpecker holes, or old tree nests. Sights: seen now in late fall and winter and after deep winter snows force it south for food because mice are hidden farther north.

Pygmy Owl *(Glaucidium gnoma)*. Length: 7-7½″; sharply outlined black patch back of ear; black streaks on brown breast; tail often held up at sharp angle. Habitat: coniferous forests, patches of young birches. Sights: flies down from tree, then straight, and suddenly swings up to new branch, like a shrike.

Pygmy Owl

Ferruginous Pygmy Owl *(Glaucidium brasilianum)*. Length: 6½-7″; like above, but streaks on breast are brownish instead of black. Different habitat: mesquite thickets, saguaros, river woods in low desert country, etc. Sights: as above.

Elf Owl *(Microthene whitneyi)*. Length: 5-6″; our tiniest owl, like thick-bodied brown sparrow. Habitat: trees of dry desert washes and saguaro cactus. Sights: flies down to attack large insect, scorpion, spider, etc., or seen peering out of hole in cactus.

Elf Owl

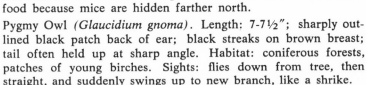

Loons, swimmers in open water with long daggerlike bills, dive almost instantly after fish under water, but have difficulty getting off surface of calm water with wings.

Loons *(Gavia species)*. Length: 23-36″; longer-bodied and larger than most ducks, thicker-billed than grebes; broad webbed feet stick out behind tail in flight. Habitat: inland and coastal waters, nests on far northern lakes. Sights: seen thrashing along surface of water to take off; diving and swimming gracefully in the roughest waves, gliding over smooth water in moonlight.

Loon

Range of Ferruginous
Pygmy Owl

Range of Elf Owl

Range of Loon

Herons and bitterns. Large birds with long legs for wading and long bills for harpooning fish while birds stand still in water; in flight head and neck form S while legs trail behind.

Great Blue Heron

Little Blue Heron

Black-crowned Night Heron

Yellow-crowned Night Heron

American Bittern

Great Blue Heron *(Ardea herodias)*. Length: 3½-4½'; great size, blue-gray color, and 6' wingspread. Habitat: in or near water, streamside woods. Sights: flying low with slow powerful strokes over marsh or stream; standing still in moonlight in water to spear fish or frogs with sudden stroke.

Little Blue Heron *(Florida caerulea)*. Length: 19-22". As above, but smaller, and with dark brownish color on neck.

Black-crowned Night Heron *(Nycticorax nycticorax)*. Length: 22-28"; comparatively stocky, short-legged, short-billed heron; black back and crown; white breast and face, gray on wings; yellow legs; red eyes. Habitat: in or near water at night. Sights: flying from nest to marshes or lakes for food; hunting while standing still in water; striking suddenly at prey, or stalking prey slowly; eyes flash red in flashlight.

Yellow-crowned Night Heron *(Nyctanassa violacea)*. Length: 22-28"; like above, but with bluish gray general color; black head, white cheek and whitish yellow crown. Habitat: swamps, marshy lakes, salt marshes. Sights: flying about on moonlit nights; hunting for crabs and crawfish by stalking them slowly in shallows.

American Bittern *(Botaurus lentiginosus)*. Length: 2-3'; wingspread: 32-50"; a large brownish bird that flies slowly, with large black streak on side of neck, black-tipped wing feathers, and long greenish yellow bill. Habitat: marshes, swamps. Sights: gives a remarkable series of neck-twistings and writhings at the same time it is producing the strange thumping or pile-driving-like sound heard in the swamp or marsh on spring nights or at dawn; seen one moment as if a dead snag, next comes alive with spearing motion of head as it strikes for live food.

Range of Great Blue Heron

Range of Little Blue Heron

Range of Black-crowned Night Heron

Range of Yellow-crowned Night Heron

Range of American Bittern

Least Bittern *(Ixobrychus exilis)*. Length: 11-14"; wingspread: 17-19"; has black back and top of head; large yellowish brown wing patches and smaller patch on side of neck; white undersides. Habitat: marshes. Sights: migrates at night, flying low and slowly; hunts prey like other bitterns and herons.

Least Bittern

Geese and ducks.

Canada Goose *(Branta canadensis)*. Length: 25-43"; black head and neck contrast with white throat and breast; otherwise gray-brown. Habitat: marshes, lakes, grasslands, grain fields. Sights: flying at night in V-shaped flocks across moon; feeding in grain fields; beating waters of lake to froth when disturbed.

White-fronted Goose *(Anser albifrons)*. Length: 27-30"; gray neck, white face, black splotches on light grayish belly are distinctive. Habitat: marshes, lakes, grasslands, grain fields. Sights: often seen feeding at night in grain fields or grasslands.

Surface-feeding Ducks (Mallard, Baldpate, Pintail, Gadwall, Teals, etc.). Length: 12-16"*. These are ducks that feed mainly close to the surface of the water. Sights: may be noticed at night when disturbed; they spring up from water directly and take wing, and feed by tipping body up so tail rises in air and head is under water.

Diving Ducks (Redhead, Ring-necked Duck, Canvasback, Scaups, Goldeneyes, Bufflehead, Scoters, etc.). Length: 12-16"†. These are ducks that dive to the bottom of ponds or rivers for food. Sights: may be noticed at night when disturbed; they take wing by paddling along on surface of water until they catch the power of flight. They feed by diving straight down.

Canada Goose

White-fronted Goose

---

* The teals are somewhat smaller, ranging from 10 to 11 inches in length.
† The bufflehead is somewhat smaller, ranging from 10 to 11 inches in length.

**Range of
Least Bittern**

Range of Canada Goose

Mallards

**Range of
White-fronted Goose**

Range of Mallard

Range of Canvasback

Canvasback

Merganser

**Range of
American Merganser**

**Range of
Hooded Merganser**

**Range of
Red-breasted Merganser**

Limpkin

Range of Limpkin

Range of King Rail

Range of Clapper Rail

King Rail

Mergansers *(Mergus* species). Length: 16-26″. Long slim bodies and toothed beaks are distinctive. Often swim and dive in swift water, and catch fish underwater at high speed. Sights: rising off water by paddling, as diving ducks do; may dive from flight at angle into water after fish.

Shore birds and marsh birds (plovers, sandpipers, etc.). Usually long-legged, long-billed birds who feed on shores and grasslands or in marshes or swamps by digging into mud or sand with bill for worms, etc., but plover bills are comparatively short. Often fly or feed at night.

Clapper Rail

Limpkin *(Aramus guarauna)*. Length: 28″; a large brown bird, white-spotted and streaked, very long dark legs; long, slightly down-curved bill. Habitats: swamps and marshes in dense growth. Sights: perched like a statue on a large tree limb, wading about in dense water vegetation, or bowing and teetering on an old root or cypress knee.

Rails, gallinules and coots. Chickenlike in appearance, though gallinules and coots more resemble ducks. Rails usually fly rather weakly, with legs dangling, dropping into marsh. Coots and gallinules rise from water after long paddling.

King Rail *(Rallus elegans)*. Length: 15-20″; large reddish yellowish brown rail, with very long, slightly down-curved bill, barred sides, and white belly. Habitats: marshes, grain fields, grasslands. Sights: flies with rapid strokes of wings low to ground; probes in mud for worms, etc.

Clapper Rail *(Rallus longirostris)*. Length: 14-17″; similar to king rail, but grayish brown. Habitats: salt marshes. Sights: sneaking silently through salt plants, or over mud at low tide, probing for worms.

Virginia Rail

Virginia Rail *(Rallus limicola)*. Length: 9-11″; like a small king rail in appearance, but with gray on side of head and shorter,

Range of
Virginia Rail

Range of Sora Rail

Range of Black Rail

Sora Rail

Range of
Purple Gallinule

Range of
Florida Gallinule

Black Rail

straighter reddish bill instead of yellowish brown. Juveniles are blackish on body. Habitats: freshwater marshes, grain fields. Sights: especially active in twilight, when it may be seen agilely climbing rushes, shrubs, vines, even small trees after grasshoppers, beetles, etc.

Sora Rail *(Sora carolina)*. Length: 8-10″; adult is grayish blue in front with black face and throat, brown above, streaked black on sides and with white rump; immature is light yellowish brown in front and without black on face. Habitats: freshwater and salt marshes; grain fields, swamps, and tidal streams. Sights: slow, weak flight for short distance, dropping into marsh; secretly slipping between grass stems; walking on bottom of pond; probing mud for small life.

Black Rail *(Laterallus jamaicensis)*. Length: 5-6″; all-black color and tiny size are distinctive. Habitats: salt marshes, wet meadows, grain fields. Sights: when closely approached, it may squat down and push head under plant cover, where, if seen, it can be caught; feeds on insects, etc. on plants.

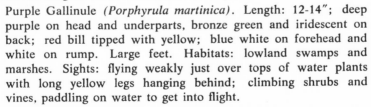

Purple Gallinule

Purple Gallinule *(Porphyrula martinica)*. Length: 12-14″; deep purple on head and underparts, bronze green and iridescent on back; red bill tipped with yellow; blue white on forehead and white on rump. Large feet. Habitats: lowland swamps and marshes. Sights: flying weakly just over tops of water plants with long yellow legs hanging behind; climbing shrubs and vines, paddling on water to get into flight.

Florida or Common Gallinule *(Gallinula chloropus)*. Length: 12-15″; general slate-gray color, white on sides and rump; red bill, yellow-tipped; large feet. Habitats: marshes, swamps. Sights: using large feet to run across water plant leaves and even on top of water; swims with head bobbing back and forth; feeds like surface-feeding duck by tipping up body while head is straight down in water grabbing plant food.

Florida Gallinule

Coot

Range of Coot

Range of
Semipalmated Plover

Range of Killdeer

Semipalmated Plover

Killdeer

Woodcock

Range of Woodcock

Coot *(Fulica americana)*. Length: 13-16"; the only ducklike bird of slate-gray to blackish color having whitish bill. Habitats: more in open water of marshes than gallinules; also lakes, ponds, swamps. Sights: swimming with head bobbing back and forth; running across water with feet beating to get into flight; diving quickly.

Plovers and turnstones. Birds that wade in shallow water and run about the shores, more stoutly built and with a thicker neck than sandpipers; often with bold patterns of color; bills generally thicker and shorter.

Semipalmated Plover *(Charadrius hiaticula)*. Length: 6-8"; dark brown above; single black bar across white chest and throat; white bar on face; bill rather small, short with black tip. Habitats: bay and ocean shores; tidal flats; salt marshes. Sights: flock breaks up when it lands, with individuals moving briskly in different directions with heads up, only to suddenly stand still, nervously bobbing and jerking their heads, then dabbling quickly at ground for worms, etc.

Killdeer *(Charadrius vociferus)*. Length: 9-11"; two black bands on white breast, plain brown back and rufous rump are distinctive. Habitats: marshes, fields, gardens, parks, riverbanks, shores, irrigated farms. Sights: at mating season particularly, killdeers may fly low overhead in the dark, dodging and swirling in marvelous evolutions; a mother, if you get near young or nest pretends to have a broken wing and to be in great pain, leading you away to a safe place; adults probe in lawns, grassy areas, etc. for small life.

Woodcock, snipe, sandpipers, etc. Bill usually longer and more slender than that of plovers, and legs usually longer; body more slender except for woodcock; most not too colorful.

Woodcock *(Philohela minor)*. Length: 10-12"; brown dead-leaf pattern of color on thick body, distinctive neckless look, very long bill. Habitats: woodland swamps and thickets, fields, streamside woods. Sights: flutters up with flight from ground when flushed; migrates in the moonlight; digs up earthworms with long bill; male flies high in the air to sing and show off by varied flight to female; coming to ground, male bows to female and struts about like a turkey cock, sometimes tripping and falling.

Wilson's Snipe *(Capella gallinago)*. Length: 10-12"; general brownish color, with light stripe over eye, grayish white undersides; distinctively long slender bill. Habitats: boggy edges of

**Range of
Wilson's Snipe**

**Range of
Spotted Sandpiper**

**Range of
Solitary Sandpiper**

Wilson's Snipe

Range of Willet

**Range of
Greater Yellowlegs**

Spotted Sandpiper

little streams and meadows, freshwater marshes. Sights: zigzag, crazy flight when flushed; male flies high in sky at mating season, circling widely and producing strange winnowing sound with wing feathers when plunging downward; male also struts in front of female with widely spread tail; plunges long bill deeply into mud after worms, etc.

Spotted Sandpiper *(Actitis macularia)*. Length: 7-8"; in breeding plumage the breast appears covered with large round spots, unique in a shore bird; in fall and winter the breast is all-white; back and head brown; white line above eye. Habitats: gravelly lake shores, ponds, streamside woods, seashores in winter. Sights: peculiar unbalanced teetering motion all the time on ground; flies with very shallow stiff wing strokes.

Solitary Sandpiper

Solitary Sandpiper *(Tringa solitaria)*. Length: 7½-9"; dark wings and back contrast with conspicuous dark- and white-barred tail; white eye-ring. Habitats: streamside woods, woody swamps, freshwater marshes, ponds. Sights: graceful, sweeping flight, especially at mating time; carefully stirs silt in pool bottom with foot to make insects rise in water and seize them; often nods to another sandpiper so solemnly that it is funny; when it alights, wings are lifted high so that tips almost touch, then lowered, and folded.

Willet

Willet *(Catoptrophorus semipalmatus)*. Length: 14-17"; large size of this gray-colored sandpiper and brilliant black and white patterns of wings are distinctive. Habitats: freshwater marshes, damp meadows, but goes to tide flats, beaches, salt marshes in winter. Sights: wings held up high when it alights; sometimes seems to bow backwards or bob up and down; flight direct with shallow flat wingbeat.

Greater Yellowlegs *(Totanus melanoleucus)*. Length: 12-15"; large slim sandpiper with long yellow legs; otherwise grayish.

Greater Yellowlegs

Lesser Yellowlegs

Semipalmated Sandpiper

American Avocet

Black-necked Stilt

Ruffed Grouse

Habitats: marshes, mud flats, ponds, streamsides; visits spruce bogs and wooded muskegs in summer. Sights: white rump flashes in flight, while dark wings curve downward, beating vigorously; holds wings high when alighting and then teeters body; when walking in shallow water, often kicks out with legs.

Lesser Yellowlegs *(Totanus flavipes)*. Length: 9-11″. Resembles greater yellowlegs, but smaller. Habitats: marshes, mud flats, ponds, shores; visits northern woods in summer. Sights: flock often moves as one through many gyrations; on shores moves about quickly and gracefully.

Semipalmated Sandpiper *(Ereunetes pusillus)*. Length: 5½-6½″; small grayish sandpiper with rather short straight bill (commonest of the small "peeps" on our shores). Habitats: shores, beaches, mud flats, etc. Sights: flocks move as units in military formations and gyrations; do much head-shaking to swallow large morsels of food; tail flicked nervously from side to side; chase each other on beach with raised wings.

Avocets and stilts. Slender wading birds with excessively long legs and long slender bills.

American Avocet *(Recurvirostra americana)*. Length: 15-20″; bill upturned; striking black and white back and wings. Habitats: marshes, alkali sinks, mud flats, ponds, bay shores (winter). Sights: catches food by swinging upcurved bill like a scythe through shallow pools while moving forward at slow run and picking up small creatures; expert swimmer.

Black-necked Stilt *(Himantopus mexicanus)*. Length: 13-17″; remarkably long red legs; body white and black; wings black. Habitats: grassy marshes, pools, lakes, mud flats. Sights: steps along high and gracefully, walking on water plants with wide feet; often tosses head.

Range of Lesser Yellowlegs

Range of Semipalmated Sandpiper

Range of American Avocet

Range of Black-necked Stilt

Range of Ruffed Grouse

Grouse and other chickenlike fowls.

Ruffed Grouse *(Bonasa umbellus)*. Length: 16-19″; reddish brown or grayish brown in general color, with a fan-shaped tail broadly black-banded and white-banded at tip. Habitats: open woods and brushlands, especially burned-over areas. Sights: whirring up out of covert when too closely approached; male strutting on log while producing deep drumming noise; female leading young and causing them to scatter and hide with warning cry; especially active on moonlit nights.

Bobwhite *(Colinus virginianus)*. Length: 8½-11″; brownish general color, but white barred with brownish black below and with reddish brown streaks along white sides; white throat and white curved bar over eye, black curved bar through eye. Habitats: brushy, open country, agricultural land, roadside hedges. Sights: flock feeding in moonlight on insects and seeds; flock whirring up from ground when disturbed; males strutting to show off to females in courting.

Bobwhite

Order of Caprimulgiformes (goatsuckers, whippoorwills, nighthawks, etc.).

Chuck-Will's-Widow *(Caprimulgus carolinensis)*. Length: 11-13″; grayish brown, with many light yellow-brown markings, including strip across throat; black streaks on head; mouth extremely wide (2″) and surrounded by hairs to catch insects and even small birds with great open mouth. Sights: flutters about when startled like a great moth, or flies up from ground after insect.

Chuck-Will's-Widow

Whippoorwill *(Caprimulgus vociferus)*. Length: 9-10″; brownish general color with white collar on throat, and white outer tail feathers on male; wings much shorter than tail. Habitats: broad-leaved woodlands, gardens. Sights: leaping from ground or flying low and erratically, to seize moths and other insects, white tail feathers of male flashing; the mother bird may flutter and tumble about as if badly hurt to draw you away from her eggs or young.

Whippoorwill

Poorwill *(Phalenoptilus nuttallii)*. Length: 7-9″; mottled brownish color, with white tips and black sides to tail; white band under chin. Habitats: chaparral, rocky hills, pinyon-juniper woodland. Sights: seen flying up from ground to seize moth or other large insect; flutters away like large moth when startled.

Poorwill

Range of Bobwhite

Range of Chuck-Will's-Widow

Range of Whippoorwill

Range of Poorwill

**Range of
Common Nighthawk**

**Range of
Lesser Nighthawk**

Nighthawks

Nighthawks *(Chordeiles* species). Length: 8-10″; streaked gray-ish brown in color, with white throat and large white blotch near middle of wing. Habitats: likes open country or open pine woods; often found in cities, uses rooftops for nests. Sights: flying through sky, from low to high, snapping up insects in enormous mouth; male may dive earthward from height and make loud booming noise to impress female; mother bird may act crazy as if badly wounded to attract enemy away from young or eggs.

Swifts. Swift and high-flying birds with very short blunt tails and very long pointed wings that seem to twinkle in flight.

Chimney Swift *(Chaetura pelagica).* Length: 5-6″; sooty brown-ish black in color; wings bowed like a scimitar in flight; flight is batlike. Habitats: open sky, especially over fields and towns; breeds in chimneys, caves, hollow trees. Sights: may chase insects high in air after dusk at high speed and especially on moonlit nights.

Chimney Swift

Perching birds. Have five toes with long claws particularly adapted to perching on twigs and branches.

Swallows. Have long, narrow wings for swift flight, but more pronounced tail than swifts, and wings do not appear to twinkle in flight.

Purple Martin *(Progne subis).* Length: 7-8½″; male distinctively all blue-black; female dark brownish black above, but white below and on neck. Habitats: open woods and lumbered or burned coniferous forests; also towns, cultivated areas, saguaro deserts; usually nesting in colonies in buildings, hollow trees, woodpecker holes in saguaro cacti, etc. Sights: seen catching insects at dusk by flying in circles with quick flapping, then gliding of wings.

Purple Martin

**Range of
Chimney Swift**

**Range of
Purple Martin**

Range of Mockingbird

Range of Catbird

Mockingbird

Mockingbirds and thrashers or mimic thrushes. Fairly large birds with strong bills, stout legs, and very long tails.

Mockingbird *(Mimus polyglottos)*. Length: 9-11″; generally gray in color, but with large white patches on wings and tail corners that show particularly well in flight; no black mask like the similar-appearing shrike. Habitats: gardens, orchards, farms, brushy areas, streamside woods (especially in deserts), etc. Sights: often seen flying or singing on moonlit nights, sometimes singing in flight above trees; often lifts up wings to display white patches on undersides.

Catbird

Catbird *(Dumetella carolinensis)*. Length: 8-9½″; a slim dark gray bird with a black cap, dark tail and reddish brown rump. Habitats: garden thickets, swampy thickets, and brushy areas. Sights: sneaks through dense foliage; male displays reddish brown rump to female, with much wheedling and strutting.

Veery

Thrushes. Chunkier bodies than the above, with large eyes, often brown-backed with spotted breasts.

Veery *(Hylocichla fuscescens)*. Length: 6½-7½″; small in size for a thrush, with general reddish brown color; spots indistinct. Habitats: moist woods and riverbottom forests. Sights: singing in moonlight from branch of tree or shrub.

Swainson's Thrush *(Hylocichla ustulata)*. Length: 7-8″; brown above; spotted breast. Habitats: streamside, woodlands. Sights: Singing at night in bushes or trees near streams.

Swainson's Thrush

Wood Thrush *(Hylocichla mustelina)*. Length: 7½-8½″. Largest and most robust of true thrushes; bright brown above, olive brown on rump and tail; whitish below; large rounded blackish

Wood Thrush

Range of Veery

Range of Swainson's Thrush

Range of Wood Thrush

Hermit Thrush

**Range of
Hermit Thrush**

spots on breast and sides; large dark eye. Habitats: moist woods and thickets; nest of leaves, grass, etc. lined with rootlets. Sights: a woodland minstrel second to hermit thrush. Sings from the higher branches of tall trees until dark night has fallen.

Hermit Thrush *(Hylocichla guttata)*. Length: 6½-7½". Brown tail and rump, olive head and shoulders (wood thrush reverses the colors). When perched, cocks its tail and drops it slowly; deep hazel eyes. Habitats: cool woods and swamps; nests on ground of moss twigs, etc. lined with rootlets and pine needles. Sights: winter months in small groups in wayside thickets and orchards. Sings during summer months and in this capacity is sometimes called the "American nightingale." No great gaps or bass notes in the hermit's song as in that of the wood thrush.

CHAPTER FOUR

Probing the Nighttime World
of Reptiles and Amphibians

Reptiles AND AMPHIBIANS include some of the most interesting creatures seen in the dark, and some of the quietest and the noisiest. Reptiles are easily told from amphibians by the hard scales that cover their bodies, the only seeming exception being the gecko lizards, whose bodies feel soft and smooth like most amphibians. However, a little close examination shows that they too have scales. All lizards have easily broken-off tails (to aid them in escaping enemies), which similar-looking salamanders do not have. All amphibians—which include the salamanders, newts, toads, and frogs—have smooth or warty skins which do not have scales and generally have to stay moist all the time or the animal shrivels up and dies. This makes it necessary for amphibians to stay near water and damp places, and, in dry times, to travel overland only at night.

How Reptiles Escape Desert Heat

Reptiles—which include alligators, crocodiles, turtles, tortoises, lizards, and snakes—show some extraordinary adaptations to living at night and for escaping extreme desert heat. Their need to escape heat was brought home to me forcibly one day in the desert town of Palm Springs during the summer of 1933, long before that place became a popular air-conditioned resort. Another naturalist and I had some snakes in cages in the shade, but several snakes suddenly went into convulsions. We immediately poured water over them, but even so a few died. Some snakes escape the desert heat by hiding in deep holes by day and hunting only at night, so developing "cat eyes," or eyes with vertical pupils and wide lenses to catch as much light as possible. Other snakes, such as the sand snakes and the shovel-nosed snakes, have little need for eyes, as much of their time is spent buried in sandy soil, or swimming through it several inches below the surface, where they feed on sand crickets and other sand-inhabiting insects. On wet nights they often come to the surface because the insects too are driven to the surface; so this is a good time to look for them.

When Reptiles Seek Heat

On the other hand, on cool nights in the spring or fall, many snakes and night-hunting lizards may be attracted to black-topped roads because the black tar of these roads holds the heat of the day much longer than the surrounding areas. By driving along such roads very slowly in a car, the nighttime explorer can pick up a snake or lizard traveling or resting on the road with his headlights and stop to examine it or follow it to find out something about its habits.

The Gila Monster and Sidewinder Rattlesnake

Two of the most extraordinary inhabitants of the desert night are the Gila monster and the sidewinder rattlesnake, both poisonous to man, but the former comparatively unlikely to be dangerous as long as you leave it alone because of its slow movements and its inability to strike as far as a rattlesnake. The sidewinder rattlesnake is dangerous to man, but its bite does not often bring death, especially if it is properly treated. The big mottled and beaded lizard travels clumsily about, almost seeming to drag its massive head and large heavy body and fat tail, till it finds a place to dig out a small rodent from the ground or finds eggs of lizards and birds. It prefers places not too far from water and is most likely to be seen traveling on wet, warm nights after rains. The sidewinder, on the other hand, is an inhabitant of dry, sandy deserts, where it holes up during the day, but moves over the sandy surfaces at night with a peculiar J-shaped movement of its body, one J up in the air while the other J is sliding over the ground, and leaving a series of J-shaped marks in the sand. Like all rattlesnakes it is usually found hunting warm-blooded mammals, whose presence it tracks by the heat-sensing pits behind its nostrils.

Identification of Reptiles
by Geographical Range and Habitat

In the Guide to Sights of Reptiles later in this chapter, many reptiles are best identified by the habitats in which they live and the geographical range, shown by the maps. Thus the desert night lizard is easily told from the similar Arizona night lizard by the fact that the latter is found in entirely separate areas from the former or in central and south Arizona. In the same way the Texas banded gecko is found in an area of Texas, Mexico, and southern New Mexico entirely separate from the common banded gecko, found more to the West. Thus, the fact that they look alike is of little importance if you know the locality in which you are looking. The granite night lizard and the desert night lizard, on the other hand, are often found in the same geographic range, but can always be told apart by the fact that the former lives around and hidden under slabs of granite or similar rock, while the desert night lizard prefers less rocky areas and particularly likes the habitat made by clumps of Joshua trees, preferably dead, and other related yucca plants.

Identification of Reptiles by Distinctive Movements

Look particularly for peculiarities of movement and of hunting when attempting to identify various reptiles. I have already mentioned such distinctive movements of the Gila monster and sidewinder, but geckos and night lizards can often easily be separated by the fact that the former are quite agile climbers, while the latter move slowly when climbing. Geckos may also leap on their prey, whereas the night lizards creep up on and suddenly bite large insects they find. Many of these differences are noted in the descriptions contained in the Guide to Sights of Reptiles later in this chapter.

Aestivation of Toads and Frogs

Amphibians are so closely connected with water and damp places that many people usually automatically expect not to see them in dry areas. However, a rain, even in the driest desert, will almost always bring them out, showing that they have been present though often deeply hidden from sight. Spadefoot toads, for example, have a special digging tool or spade on the hind foot that enables them to dig a deep burrow in which they can hide and even hibernate or aestivate during a long dry period in desertlike regions. As a pond dries up, other toads and frogs bury themselves in the mud at the bottom, and stay alive by aestivation (a kind of long sleep) in the slight moisture that remains in the drying slabs of clay-mud. When rains revive the pond, out they come to breed and produce new frogs and toads.

How Salamanders Conserve Body Moisture

Salamanders do not have the ability to continue to exist long in dry areas, and so are mainly found away from the deserts and most often in areas of considerable plant cover and moisture. But they too hide in crevices or under piles of brush and wood during the dry seasons and come out again with the rains. They also do a lot of hiding under rocks, boards, and trash during the daytime and are most active hunting for insects and worms in the comparative cool and damp of the night. Some kinds are more likely to be found far from permanent water than frogs, and some are not dependent on bodies of water to have their young, but, as do the lungless salamanders (Plethodontidae), have the young come out of the eggs fully formed and ready for terrestrial life without a period of water-living. On the other hand, some salamanderlike animals—the efts, mud puppies, and sirens, for example—rarely if ever leave the water; and some spend all their lives in the gill-breathing, larvalike state.

Best Times and Places to Hunt for Amphibians

A night following a good warm rain is a wonderful time to watch and look for activity among amphibians. Not only do great numbers come out to hunt at such times, but there may be a frenzy of activity to prepare for and take part in mating, particularly in places near which or in which any water has gathered. On the Pacific coast extraordinary number of newts, even covering the forest floor, move overland towards water on such nights, each ready, as soon as touched or frightened, to twist its body into a strange half-circle, exposing the bright orange underside that warns all predators that it is dangerously poisonous when eaten.

Many other amphibians have similar actions or habits that help differentiate them when seen. The seal salamander may be seen poised on a wet rock in the forest with the poised upright posture of a seal, alarmed by an enemy's approach and about to dive into the ocean waves. The red-legged frog and other similar long-legged frogs hunt for insects along the banks of streams or ponds, but are ready in an instant to take a tremendous leap down into or near the water, immediately diving to the bottom to hide among the water plants. Narrow-mouthed toads, on the other hand, are too short-legged for much leaping, but can run or dart very rapidly into nearby cover and holes when people or animals try to catch them. Gopher and crawfish frogs are similar poor jumpers, but usually stay in water or very close to their tunnellike hiding places.

Mating Habits of Amphibians and Reptiles

The mating habits of both amphibians and reptiles are often extraordinary and best seen at night when the mating frenzy seems to reach its height. Some salamander males, in the water, coil and lash about the body of the female, and then deposit a cone-shaped capsule of sperm, which the female picks up with the lips of her cloaca and sucks into her body. Painted turtle males, on the other hand, have extraordinarily long nails on their front feet which they use to tickle the neck and head of the female while swimming beside her and so induce her to be courted. Most frog and toad males wait in water for their females, calling them by an absolutely amazing variety of sounds, then seizing them from behind in the mating grip, a grip so powerful that it even persists after death!

Identification of Amphibians
by Habitat and Geographical Range

Amphibians, like the reptiles, are often best separated from each other by studying the habitats in which they live or their geographical ranges. Thus you can use the maps in the Guide to Sights of Amphibians later in this chapter to learn such facts as that the spadefoot toads are all found in comparatively dry areas of the country, grasslands or deserts, while all frogs are found only near or in permanent water. The dusky salamanders and the woodland salamanders look much alike, both kinds being rather slender and both inhabiting similar territory; but the woodland salamanders are always found in and around rocky or woody areas, while the dusky salamanders are associated mainly with springs, brooks, and swift streams, thus making it easy to tell the difference between them.

If different amphibians and reptiles are seen in night explorations, carefully compare their shapes with those shown in the silhouettes contained in the Guide to Sights of Reptiles and the Guide to Sights of Amphibians below, watch their habits for those described in the Guides, and look for proper habitats and geographical ranges before you decide upon the identity of the creatures which have just been seen. An expert at a university or museum can often help to confirm an identification.

Because most reptiles and amphibians are well hidden and hard to see at night, and when seen are hard to differentiate without more technical detail, they are listed in the Guides mainly by their generic groups, except for some species that are particularly common or easy to identify. Chapter 9 on sounds, however, gives the specific sounds of most all of the species, and these combined with the maps of the genera should often help in identifying exactly what kinds are being heard. For further help in sight identification of reptiles, refer to the two field guides to reptiles and amphibians described in the Selected References at the back of this book.

GUIDE TO SIGHTS OF REPTILES

(Note: The dimensions given in this Guide are those of length. Length in crocodiles, turtles, snakes, and tortoises is total length; length in lizards is minus the tail.)

Crocodilians, family Crocodylidae: crocodiles, caimans, and alligators.

American Crocodile *(Crocodylus acutus).* 7-15'; gray to brownish gray; narrow snout; large tooth in lower jaw is visible even when mouth is closed. Habitats: swamps of southern Florida. Sights: resting in water partly submerged; sliding up on bank or down into water.

American Alligator *(Alligator mississipiensis).* 6-19'; usually blackish; snout rather broad and shorter than crocodile; no tooth showing when mouth closed. Habitats: river swamps, lakes, marshes, etc. in South. Sights: lying partly submerged in water or swimming slowly; moving up on bank; eyes shine ruby red in light; may snap up animal that comes down to drink.

Ranges of
American Crocodile
and American Alligator

Turtles and tortoises. All have shell of some kind.

Snapping turtles. Have large ugly heads, with strong shearing jaws.

Snapping Turtle *(Chelydra serpentina).* 8-20"; 10-80 lbs. or more; typical small back shield and long sawtooth tail identify it. Habitats: almost anywhere in fresh water, sometimes in brackish. Sights: feeding on vegetation or hunting small life on shore at night.

Snapping Turtle

Alligator Snapping Turtle *(Macroclems temmincki).* 15-35" or more; 35-200 lbs.; head huge and back shield extremely rough-ridged. Habitats: same as snapping turtle. Sights: rather rare, but may be seen swimming at night in moonlight.

Alligator
Snapping Turtle

Range of Alligator
Snapping Turtle

Range of
Snapping Turtle

Stinkpot

**Razor-backed
Musk Turtle**

**Loggerhead
Musk Turtle**

**Stripe-necked
Musk Turtle**

Flattened Musk Turtle

Striped Mud Turtle

Sonora Mud Turtle

Musk and mud turtles. Musk turtles have small plastrons (bottom shields) hinged in only one place, giving little leg protection. They are known by their strong musky smell when handled. The mud turtles have a good deal larger plastrons, which are hinged at both ends in two places.

Stinkpot (*Sternothaerus odoratus*). 3-5"; only musk turtle with two light stripes on head and two barblike projections on chin. Habitats: generally in large sluggish streams. Sights: may fall into your boat at night if you brush a small tree on the side of watercourse where turtle is sleeping; sometimes seen walking on bottom in clear, shallow water.

Razor-backed Musk Turtle (*Sternothaerus carinatus*). 4-5½"; distinctive single keel on back. Habitats: streams and rivers swamps. Sights: may climb out on log on moonlit night.

Loggerhead Musk Turtle (*Sternothaerus minor minor*). 3½-5"; very large head and distinctive three keels on back. Habitats: springs in Florida. Sights: may be seen climbing in small tree, bush, etc.

Stripe-necked Musk Turtle (*Sternothaerus minor peltifer*). 3-4½"; has many faint stripes on neck. Habitats: likes clear streams and creeks. Sights: walking on bottom.

Flattened Musk Turtle (*Sternothaerus depressus*). 3-4⅛"; distinctive flattened appearance. Habitats: Alabama rivers and tributaries of Black Warrior River.

Striped Mud Turtle (*Kinosternon bauri*). 3-4½"; usually has three light stripes on back shell, and two light stripes on each side of head. Habitats: likes shallow water, wet meadows, ditches, etc.; often found on land at night. Sights: crawling about in grass.

Other Mud Turtles (*Kinosternon* species in part). 3-5"; backs not striped. Habitats: usually in muddy water of sloughs, marshes, ponds, sometimes in wet meadows or woodlands. Sights: seen

Range of Stinkpot

**Ranges of Loggerhead
and Stripe-necked
Musk Turtles**

**Ranges of
Razor-backed and
Flattened Musk Turtles**

**Ranges of
Striped Mud and
Western Pond Turtles**

**Range of Mud
Turtles Without
Stripes on Backs**

feeding on grass in meadows at night. Sonora mud turtle *(Kinosternon sonoriense)* feeds in woodlands at night in grassy clearings.

Water turtles, family Testudinidae, subfamily Emydinae. These are commonly seen water turtles, with webbed toes.

Western Pond Turtle

Western Pond Turtle *(Clemmys marmorata)*. 3-7"; the dark-colored carapace (back shield) usually has each scute with faint or dark lines radiating from center. Habitats: streams, rivers, marshes, ponds, usually where there is plenty of aquatic vegetation. Sights: sometimes gets out on log in moonlight; found on land near water; sometimes travels to lay eggs in sandy soil.

Spotted Turtle *(Clemmys guttata)*. 3½-5"; few to fair number of yellow spots on dark back, but not crowded. Habitat: marshes, marshy meadows, bogs, ponds, swamps, etc. Sights: leisurely moving over land near water at night while searching for plant food.

Spotted Turtle

Wood Turtle *(Clemmys insulpta)*. 5½-9"; shell very rough, each scute carved in pyramidal form; orange red on neck and forelegs. Habitat: shallow water, but also meadows, woods, farmlands, etc. Sights: commonly seen wandering on land at night in search of plant food. Makes excellent pet.

Box turtles, subfamily Emydinae. Always told by the plastron, which is so hinged that body, tail, and head can be drawn inside and completely protected.

Wood Turtle

Ornate Box Turtle *(Terrapene ornata)*. 4-5½"; beautiful radiating dark lines on each shield against a yellow background. Habitats: plains, prairies, especially in sandy areas, open woodland. Sights: wandering about at night after plant food and insects; may tear up cow dung to get at insects inside. Warm rain brings it out in large numbers.

Eastern Box Turtle *(Terrapene carolina)*. 4½-6½"; distinctive high-domed shell, variously bright and dark-colored, sometimes radiating streaks. Habitats: open woods, grasslands, meadows, brushy areas. Sights: wandering around at night feeding on plants, insects, etc.; may be especially plentiful after a warm rain.

Ornate Box Turtle

Terrapins, map turtles, and sawbacks. Generally with concentric rings, maplike areas, or round spots on each scute of the carapace.

Diamondback Terrapin *(Malaclemys terrapin)*. 4-8½"; characteristically spotted or flecked on head and legs; each scute on carapace with concentric light or dark markings on ridges or

Eastern Box Turtle

Range of Spotted Turtle

Range of Wood Turtle

Range of Ornate Box Turtle

Range of Eastern Box Turtle

Diamondback Terrapin

**Range of
Diamondback Terrapin**

**Range of Map and
False Map Turtles**

**Ranges of
Some Sawback Turtles**

Painted Turtle

Slider

Map Turtle

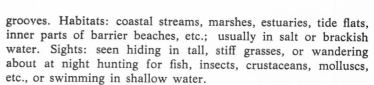

grooves. Habitats: coastal streams, marshes, estuaries, tide flats, inner parts of barrier beaches, etc.; usually in salt or brackish water. Sights: seen hiding in tall, stiff grasses, or wandering about at night hunting for fish, insects, crustaceans, molluscs, etc., or swimming in shallow water.

Map Turtles and False Map Turtle (*Graptemys* species). Most specimens have maplike markings on back. Males: 3½-6"; females: 5-10½"; usually with thorn- or knoblike projections on back. Habitats: lakes and rivers. Sights: may be seen in moonlight or dusk resting on logs, banks, or slanting tree limbs out of water; or eating snails, insects, etc. in water, or sliding into water.

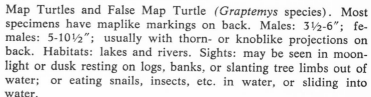

False Map Turtle

Sawback Turtles (*Graptemys* species). Most specimens have appearance of sawtoothed back or are very spiny on back. Males: 3-4"; females: 4-8". Habitats: rivers and large streams. Sights: eating insects and molluscs in water.

Painted turtles, cooters, and sliders. Either smooth, or carapace covered with wrinkles.

Painted Turtles (*Chrysemys* species). 4½-9"; beautiful red, yellow, and black patterns on back, though sometimes black almost absent; very smooth shells. Habitats: in shallow water where there is much aquatic vegetation and bottoms are muddy; ponds, ditches, sloughs, even cattle troughs, etc. Sights: may come out on land or water in moonlight on hot nights to feed on plants, insects, crayfish, and mulluscs.

**Range of
Painted Turtles**

**Range of
Cooters and Sliders**

Cooters and Sliders (*Pseudemys* species). 5-13"; fairly large turtles, with carapace toothed on rear margin and covered with wrinkles or furrows; each scute usually has whorls and streaks of black or brown against a lighter brown or olive color; face and neck usually streaked or patched with yellow or red. Habitats: found in water almost everywhere. Sights: some may come out at night to feed in grassy meadows or flats near water, or may be seen in small ponds or even cattle tanks if approached cautiously.

**Ranges of
Three Species of
Gopher Tortoises**

**Range of Spiny
Soft-shelled Turtles**

**Ranges of
Smooth and Florida
Soft-shelled Turtles**

Gopher Tortoise

Gopher Tortoises *(Gopherus* species). Hind feet almost elephant-like; forelegs heavily scaled; the carapace is high and rounded. 5½-14"; three species can be identified by location as shown on map. Habitats: plains, deserts, rocky slopes, dunes, desert oases. Sights: seen hunting plant food at night, clumsily traveling about, mainly on very warm nights when day is too hot, especially after warm rains; sometimes seen digging deep burrows in the ground.

Soft-shelled Turtle

Soft-shelled turtles *(Trionyx* species). Males: 5-11"; females: 7-18"; look almost like a pancake that can run or swim very fast; shell leathery and soft, without scales. Habitats: rivers, streams, lakes, ponds; likes muddy or sandy bars. Sights: may be seen on the bars on hot moonlit nights, or swimming rapidly in water.

Lizards. Scaly skin differentiates lizards from smooth-skinned salamanders; tails often break off when touched, which happens to legless lizards, showing they are not snakes; lizards also have movable eyelids. Most lizards are active in daylight; only a few come out at night.

Geckos, family Gekkonidae. Soft-skinned, with fine granular scales; night hunters of insects. Geckos communicate by chirping and squeaking.

Banded Geckos *(Coleonyx* species). 2-3"; dark bands across yellowish brown back; no toe pads; eyes with visible lids. Habitats: generally around or in rocks in desert areas. Sights: most often seen on black-top roadways at night if approached cautiously, but may also be seen hunting insects among rocks.

Banded Gecko

Leaf-toed Gecko *(Phyllodactylus xanti)*. 2-2⅗"; large eyes have no visible lids; toes have enlarged pads for climbing; generally brown or gray-brown. Habitats: usually in rocky canyons near streams or springs. Sights: climbing over rocks in darkness.

Reef Gecko *(Sphaerodactylus notatus)*. 1-1⅛"; very tiny brown lizard with rather thick body, brown in color and either striped or spotted. Habitats: subtropical woods, gardens, and homes. Sights: scurrying around in dark hunting insects, etc.

Leaf-toed Gecko

**Range of
Banded Geckos**

**Range of
Leaf-toed Gecko**

Range of Reef Gecko

Reef Gecko

Mediterranean Gecko

Mediterranean Gecko *(Hemidactylus turcicus)*. 2-2¼″; very big eyes; long and broad toe pads for climbing walls, etc.; covered with small bumps above; yellowish brown with rows of dark spots. Habitat: gardens and buildings. Sights: climbing over walls and screens at night; hunting insects near lights; has mouselike squeak.

Night lizards, family Xantusiidae. Slender, small, and secretive; eyes with vertical pupils; no eyelids.

Granite Night Lizard

Granite Night Lizard *(Xantusia henshawi)*. 2-2⅘″; body flat and skin soft and pliant; large dark spots on pale brownish back. Habitat: rocky areas in desert or semidesert regions; especially likes granite outcroppings in canyons near water. Sights: crawling over rocks hunting insects at night.

Arizona Night Lizard

Arizona Night Lizard *(Xantusia arizonae)*. 1¾-2¼″; the satiny skin of the olive gray to dark brown back is covered with black specks. Habitat: among rocks of oak woods, chaparral and desert scrub. Sights: crawling over rocks on warm nights after insects.

Desert Night Lizard

Desert Night Lizard *(Xantusia vigilis)*. 1½-2″; slimmest and smallest of night lizards; satiny gray-olive to dark brown skin covered with black specks. Habitat: desert scrub and rocks; frequent near or under dead Joshua trees and other yuccas or agaves. Sights: crawling about yuccas on warm nights hunting insects.

California legless lizards, family Anniellidae. Snake-like, but have movable eyelids; and the tails break off.

California Legless Lizard

California Legless Lizard *(Anniella pulchra)*. 4½-7″; yellowish below; silver-colored to light grayish brown or dark brown above; black line down middle of back. Habitats: burrows in sandy soil where there is good moisture and plant cover of beaches, chaparral, oak and pine woods, streamside woods. Sights: wriggles out at night to hunt for insect food and worms, over ground and under plant cover.

Venomous lizards, family Elodermatidae (not dangerous if left alone; protected by law).

Gila Monster

Gila Monster *(Heloderma suspectum)*. 12-16″; large heavy body and head and swollen tail are distinctive; body brightly patterned with pink, black, orange, or yellow. Habitats: Desert washes, shallow canyons, oak woods, usually near permanent or at least intermittent streams; may hide in rocks, dense thickets, and

Range of Mediterranean Gecko

Range of Granite Night Lizard

Ranges of Arizona and Desert Night Lizards

Range of California Legless Lizard

Range of Gila Monster

wood rat nests. Sights: wandering clumsily over land at night looking for small mammals, bird eggs, etc., especially after a rain.

Snakes. The great majority of snakes are not dangerous.

Slender blind snakes, family Leptotyphlopidae. Very slender, wormlike, with very tiny vestigial eyes.

Slender Blind Snakes *(Leptotyphlops* species). 6-16"; spine on tail; body of various color but with silvery sheen. Habitats: desert scrub, brushy hills, canyon bottoms, washes, sandy deserts. Sights: burrowers, but come to surface to hunt insects, worms, etc. at night, especially warm, humid nights.

Slender Blind Snake

Boas, family Boidae. Heavy-bodied snakes with glossy, smooth scales; head small.

Rosy Boa *(Lichanura trivirgata).* 22-42"; thick-bodied with head about as wide as neck; most scales on head and chin small; variously colored, but has three broad brown stripes on back, or patches in row. Habitats: rocks, brush, desert scrub, desert streamside-woodland. Sights: climbing in bushes, hunting mice and rats to constrict them.

Rosy Boa

Common Snakes (colubrids). Usually with large, even-sided head and chin plates.

Water Snakes *(Natrix* species). 24-60"; usually told by colorful and distinctive patterns of underside, two columns of scales past the tail, and bad-smelling musk given off from glands under the tail when attacked or captured. They do not have the pit behind the nose like the poisonous cottonmouth moccasin, its long fangs, or the cottony appearance of its wide-open mouth; also, they move much faster. Habitats: most water areas and their banks. Sights: seen sliding off snags, logs, rocks, etc. into water when approached, or swimming on surface.

Water Snake

Lined Snake *(Tropidoclonion lineatum).* 8-20"; identified by keeled scales, conspicuous black half-moon spots down pale belly, and single anal plate. Habitats: prairies, open woods, city dumps and parks, vacant lots, etc. Sights: hunting over ground for earthworms on damp nights.

Striped Swamp Snake *(Liodytes alleni).* 13-25"; brown and shiny water-loving snake, with a wide yellow stripe along the lower side of the body; belly yellowish to orange-brown; small head. Habitats: swamps, marshes, bay borders, bogs. Sights: seen in

Lined Snake

Range of Western Blind Snake

Range of Texas Blind Snake

Range of Rosy Boa

Range of Water Snakes

Striped Swamp Snake

Ranges of Lined and
Striped Swamp Snakes

Range of Spotted
Leaf-nosed Snakes

Range of Saddled
Leaf-nosed Snakes

Leaf-nosed Snake

Fox Snake

Glossy Snake

twilight on rainy or damp evenings on land or roads near swampy areas, also swimming in ponds.

Leaf-nosed Snakes *(Phyllorhynchus* species). 12-20″; snout appears covered with leaflike appendage that curves upward; either spotted (spotted leaf-nosed) or with saddles of brown on back (less than seventeen on body of saddled leaf-nosed). Habitats: desert scrub, usually burrowing in sandy or rocky soil during day. Sights: on roads or on soil at night hunting geckos and large insects and their relatives.

Rat, Corn, and Fox Snakes *(Elaphe* species). 24-84″; usually large snakes, always told by shape of body being like loaf of bread, whereas nearly all other snakes are like a half-moon; often very colorful with reddish or brown blotches, or with patches of dark stripes on back and side. Habitats: pine woods, farm woodlots, wooded and rocky canyons, caves, prairies, streamsides, dunes, marshes, river swamps, farm fields, old buildings, saw grass. Sights: out on ground on warm, damp nights, climbing trees and rocks, after rats, mice, birds, etc.; vibrate tails if alarmed.

Glossy Snake *(Arizona elegans)*. 26-56″; faded in color on top, with light tan or gray blotches; scales glossy and smooth. Habitats: chaparral, grasslands, desert scrub, sagebrush, some woodlands. Sights: actively chases lizards, smaller snakes, mice, and rats at night.

Gopher (in West), Pine and Bull Snakes (in East) *(Pituophis melanoleucus)*. 36-100″; our largest and most powerful snake, but tames easily; yellowish or cream-colored, with numerous brown, black, or reddish brown blotches on back, wider spaced at tail end; often mistaken for rattlesnake, especially when it vibrates its tail in dry leaves, but has head about as narrow as neck, not wide as in rattlesnake. Habitats: almost all land

Gopher Snake

Range of Fox Snakes

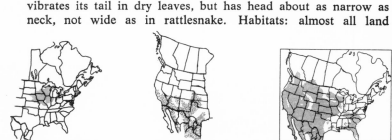

Range of
Glossy Snake

Range of Gopher,
Pine, and Bull Snakes

Range of Milk Snake

Ranges of California Mountain and Eastern King Snakes

Range of Common King Snakes

California Mountain King Snake

Milk Snake

Eastern King Snake

Scarlet King Snake

Long-nosed Snake

habitats except high mountain areas. Sights: mostly a day hunter, but comes out on warm nights to hunt gophers, squirrels, rabbits, mice, rats, often following them right into their burrows.

California Mountain King Snake *(Lampropeltis zonata)*. 20-40"; beautiful red, black, and white rings, with the black bordering both the others. Habitats: most coniferous forests, woodlands and chaparral, streamside woods. Sights: comes out at night to hunt over ground for small mammals, lizards, small birds, and smaller snakes when weather is very warm.

Milk Snake *(Lampropeltis triangulum)*. 15-54"; the rings of orange, red, or reddish brown are bordered by black rings, which are separated by yellow or white rings; also many other variations of this snake in the East, some with blotches instead of rings. Habitats: streamside woods, coniferous forests, rocky hills, prairies, sand dunes, cultivated lands, etc. Sights: comes out at night in warm weather to hunt lizards, snakes, mice, birds, and eggs. Falsely accused of milking cows!

Eastern King Snake *(Lampropeltis getulus)*. 35-75"; shiny black to dark brown, variably marked, but common form with a sharply marked chainlike pattern of cream or whitish; sometimes all black. Habitats: streamside woods, swamp borders, pine forests, mountain meadows. Sights: comes out at night in hot weather to hunt water snakes, and eggs of turtles in particular.

Scarlet King Snake *(Lampropeltis doliata)*. 14-26"; very beautiful small snake with wide red bands bordered by narrow black bands, and these with yellowish bands on other side; red snout (unlike poisonous coral snake, which has black nose and red rings bordered by white or yellow). Habitats: under bark, logs, etc. in pine forests or broad-leaf woodlands. Sights: comes out at night to hunt small snakes, lizards, juvenile mice, earthworms, and small fish.

Long-nosed Snake *(Rhinocheilus lecontei)*. 22-40"; red- and black-blotched snake with speckled appearance, the black blotches rimmed with yellow. Habitats: desert scrub, prairies, brushlands, irrigated desert areas. Sights: often wriggling on or over blacktop roads at night, or seen near irrigated areas capturing insects, mice, and lizards and their eggs.

Ground Snakes *(Sonora species)*. 8-21"; small snakes with yellowish white undersides and a reddish orange back, sometimes banded or blotched with black; head small. Habitats: sandy

Range of Scarlet King Snake

Range of Long-nosed Snake

Ground Snake

Shovel-nosed Snake

Banded Sand Snake

Hook-nosed Snake

Black-headed Snake

soil of arid or semiarid regions, sagebrush, desert scrub, thickets of willows and mesquites near surface or subsurface water. Sights: comes out at night to wriggle about looking for spiders, insects, centipedes, etc.

Shovel-nosed Snakes *(Chionactis* species). 10-18″; with shovel-shaped nose and deeply undershot jaw; yellowish white with black and red blotches or bands on back. Habitats: desert scrub, dunes, sandy or gravelly areas. Sights: comes out at night after insects, spiders, etc.; tries to burrow under and swim through sand or gravel when attacked.

Banded Sand Snake *(Cilomeniscus cinctus).* 7-10″; flat snout, eye very small; lower jaw much inset; black bands on reddish yellow body. Habitats: sandy or gravelly soil. Sights: swims under and through sand with undulating motion, but may be seen on surface at night hunting centipedes, sand-burrowing cockroaches, etc.

Hook-nosed Snakes *(Ficimia* species). 7-12″; definite upturned hook to nose; with brown or yellowish brown crossbands against lighter colors. Habitats: semiarid grasslands, woodlands, and brush; canyon bottoms, mesquite, etc. Sights: seen catching spiders, etc. on ground surface on warm nights after rains; may writhe body violently when attacked, making a popping, bubbling sound with vent.

Black-headed Snakes *(Tantilla* species). 7-18″; small slender snakes with light tan bodies, black heads, and white rings around necks. Habitats: desert scrub, mesquite deserts, rocky areas, desert streamside woods; prairies, desert grassland, oak woods. Sights: comes out at night to wriggle over ground after insects, spiders, etc.

Range of Ground Snake

Range of Western Ground Snake

Range of Western Shovel-nosed Snake

Range of Sonora Shovel-nosed Snake

Range of Banded Sand Snake

Ranges of Western and Desert Hook-nosed Snakes

Range of Mexican Hook-nosed Snake

Ranges of Black-headed Snakes

Lyre Snakes *(Trimorphodon* species). 24-42″; distinctive broad head and slender neck with mark like a lyre; verticle pupils show "cat eyes." Habitats: rocky areas in desert scrub, where they hide by day in crevices; desert grassland, oak woods, pine forests; rocky canyons. Sights: come out at night to catch lizards, mice, bats; killing them by means of poison in hind teeth (of little danger to man); may be found on blacktop roads.

Lyre Snake

Night Snake *(Hypsiglena torquata).* 12-26″; gray or grayish brown with dark gray or brownish spots; often a pair of large dark blotches on back of neck; head flat; eyes with vertical pupils. Habitats: short grass plains, sagebrush, chaparral, desert scrub, oak woods, streamside woods, rocky and sandy areas. Sights: out on warm nights, hunting lizards and frogs, which it kills with poison in back teeth; sometimes found crawling on blacktop roads.

Night Snake

Cat-eyed Snake *(Leptodeira septentrionalis).* 17-38″; distinctive bold broad dark brown to black saddles on yellowish or yellow brown ground color; vertical pupils in eyes. Habitats: streamside woods. Sights: hunting frogs near water and killing them with poisonous back teeth.

Cat-eyed Snake

Coral snakes, family Elapidae. Small fanged poisonous snakes.

Arizona Coral Snake *(Micruoides euryxanthus).* 15-22″; brilliantly colored with broad rings of red and black narrowly separated by white or yellow rings, all encircling body. Habitats: desert scrub, chaparral, oak woods, grasslands, cultivated lands. Sights: comes out at night to hunt small animals. Poison very dangerous, but rarely used on man.

Arizona Coral Snake

Vipers, family Viperidae. Pit vipers, subfamily Crotalinae.

Western Diamondback Rattlesnake *(Crotalus atrox).* 30-90″; body light grayish, often with dark diamond-shaped blotches on back, but shapes vary, and generally speckled appearance is distinctive. Habitats: desert scrub, desert grassland, brushy areas, rocky canyons, oaks and streamside woods, etc. Sights: fiercely defends itself by rattling and striking when disturbed (very poisonous); slides over ground quietly on heat-trail of rabbits, rats, etc.; detects their presence by special pits on nose.

Western Diamondback Rattlesnake

Sidewinder *(Crotalus cerastes).* 17-32″; light-colored without distinctive pattern; hornlike points on head. Habitats: sandy deserts. Sights: peculiar sidewinding motion when traveling over sandy areas at night, leaving parallel J-shaped marks in sand,

Range of Western Diamondback Rattlesnake

Ranges of Lyre Snakes

Range of Night Snake

Range of Cat-eyed Snake

Range of Arizona Coral Snake

Sidewinder

Range of Sidewinder

Range of
Black-tailed
Rattlesnake

Range of
Tiger Rattlesnake

Black-tailed
Rattlesnake

Tiger Rattlesnake

with the hook end of the J pointing out the direction the snake is taking; hunts kangaroo rats, pocket mice, etc.

Black-tailed Rattlesnake *(Crotalus molossus)*. 27-50″; tail and often the nose black, sharply contrasting with rest of body. Habitats: rocky areas in mountains and canyons; desert scrub, brushy areas, pinyon-juniper woods, and coniferous forests. Sights: active on warm, damp nights, when it hunts small mammals; not very aggressive.

Tiger Rattlesnake *(Crotalus tigris)*. 17-36″; irregular greenish gray or brownish crossbands on gray to bluish gray or light brown back. Habitats: rocky areas of canyons, foothills and desert mountains, from desert scrub to oak woodland. Sights: seen hunting small mammals at night, especially after warm rains.

GUIDE TO SIGHTS OF AMPHIBIANS

(Note: All dimensions refer to length.)

Salamanders, newts, etc. Look like lizards, but usually have gills when young, and skin is not scaly, but smooth, generally moist or slippery. Length as given is from nose to base of tail.

Mud puppies, water dogs, and sirens, family Proteidae. Both adults and young have gills.

A Salamander of the genus Necturus

Range of Necturus salamanders

Mud Puppies and Water Dogs *(Necterus* species). 8-16″; dark reddish brown plumes of the gills are distinctive as they wave languidly in current; four toes on each foot; dark stripe across eye. Habitats: most permanent bodies of fresh water. Sights: hunting at night in water for small fishes and their eggs, crayfish, water insects, snails, etc.

Sirens *(Siren* species). 7-36″; eellike salamanders with only front pair of legs present. Habitats: shallow, usually weed-choked, water areas. Sights: often continue to hunt at night in flashlight's beam for crayfish, snails, worms, and aquatic vegetation.

Amphiuma *(Amphiuma means)*. 13-40″; two pairs of tiny, useless-looking legs; large eellike body. Habitats: shallow water areas, often weed-choked. Sights: sometimes wriggles overland on wet nights; seen in water hunting worms, insects, snails, fishes, snakes, frogs, etc.

Siren

Range of Lesser Siren

Range of Greater Siren

Range of Amphiuma

Amphiuma

Range of Tiger Salamander and Relatives

Range of Pacific Giant Salamander

Range of Woodland Salamanders

Pacific Giant Salamander

Mole salamanders, family Ambystomidae. Usually have broad heads, small eyes, laterally flattened tails, and rather deep side grooves; teeth form fairly straight row (broken sometimes) across roof of mouth.

Tiger Salamander and Relatives *(Ambystoma* species). 2-7″; not with large head like Pacific giant salamander, or with square end of body like Olympic salamander; usually colorful and well marked or spotted. Habitats: Hide most of time in holes, under large objects, or in debris, usually near vegetation. Sights: come out on nights of first heavy rains to breed, some traveling in numbers overland.

Tiger Salamander

Pacific Giant Salamander *(Dicamptodon ensatus).* 4-6¼″; distinctive large massive head, dark marbling on purplish brown skin. Habitats: in or near cold streams of damp coniferous forests. Sights: feeds on frogs and salamanders; climbs small trees up to 8′; crawls into nearby damp woods; makes rattling sound.

Lungless salamanders, family Plethodontidae. Lungless, breathing through skin, which is slippery-smooth; very narrow groove extends from nose to top of upper lip; no aquatic larval stage.

Woodland Salamander

Woodland Salamanders *(Plethodon* species). 1¼-5″; body slim and legs short; stripe down back usually of yellowish-brown, yellow, reddish color, but sometimes overall black or brown. Habitats: generally in damp forests, where they hide during day. Sights: often seen in numbers in light on damp, warm nights and after rains; crawl after insects, worms; may swim.

Dusky Salamander

Dusky Salamanders *(Desmognathus* species). 1¼-5″; bodies huskier than those of woodland salamanders; hind legs larger and more powerful than front legs; usually have pale line from base of jaw to eye; much better jumpers than most salamanders. Habitats: in or near small streams, falls, springs, seeps, and nearby damp forests. Sights: on wet nights may be seen wandering over ground after worms, insects, etc. The seal salamander

Range of Dusky Salamanders

Arboreal Salamander

Range of
Arboreal Salamander

Ranges of
Web-toed Salamanders

Range of Yellow
Brook Salamanders

Web-toed Salamander

Yellow Brook
Salamander

Spadefoot Toad

(Desmognathus monticola) appears like a tiny seal when poised on top of a wet rock in the light of a flashlight; some climb up trunks of trees in night; may be seen walking slowly and deliberately along bottom of clear streams; seen jumping to catch food or when disturbed.

Arboreal Salamander *(Aneides lugubris)*. 2½-4"; brown color, usually speckled with light yellowish spots; head quite large and triangular-shaped; tail coiled at rest. Habitats: oak woodland and ponderosa pine forests. Sights: seen at night climbing tree trunks or rocks; may squeak.

Web-toed Salamanders *(Hydromantes* species). 1¾-2¾"; distinctive flat body, webbed feet, and strange, free-moving, mushroom-shaped tongue that can be flung out to catch insects at least ⅓ of body length; generally brownish in color, often with faint dark markings. Habitats: damp, rocky areas and caves. Sights: climbing steep rock faces, often moving prehensile tail forward and curling it to get grip to help climb; catching insects, worms, etc. with flick of long tongue.

Yellow Brook Salamanders *(Eurycea* species). 1-5½"; usually with some yellow color, at least on the undersides; some with dark lines down their sides; most with comparatively long thin narrow tails, some even enormously long. Habitats: mostly brooks and brookside woods, springs, caves, riverbottom swamps. Sights: jumping into water and swimming away swiftly; wandering on land hunting insects, worms, etc. on damp nights; climbing steep rock faces, using prehensile tail to help.

Toads and frogs.

Spadefoot toads, family Pelobatidae. Toads that dig holes in ground with dark spadelike projection or tubercle on each hind foot; skin comparatively smooth, and pupil of eye vertical like cat's.

Range of
Spadefoot Toads

Spadefoot Toads *(Scaphiopus* species). 1¼-3½"; generally greenish or grayish with darker markings. Habitats: wooded areas, but sandy or loose soils; savannas, plains, mesquite areas, alkali flats, sagebrush, pinyon-juniper woods. Sights: congregating at temporary water after rain to sing; hopping about hunting insects, spiders, etc. on damp, warm nights, usually after rains; digging hole in ground by working backward and rotating spades in hind feet to dig.

Range of True Toads

Range of Tailed Frog   Range of Barking Frog

Tailed Frog

True toads, family Bufonidae. Noted for dry, warty skin, short, rather clumsy hops, and thick bodies; they have no spades on hind feet like above toads; pupils of eyes round and not vertical. Poison in glands on back usually discourages predators.

> True Toads (*Bufo* species). ¾-9"; colors range from brown through gray, greenish, and yellow, often with darker markings and mottlings. Habitats: most habitats from deserts to damp woodlands and streamsides. Sights: clumsily hopping about on damp nights; coming to lights to flick insects into mouth with long hinged tongue; gathering in pools or marshes, etc. for singing and breeding.

Barking Frog

Tailed frogs, family Ascaphidae. Strange taillike organ at end of body.

> Tailed Frog (*Ascaphus truei*). 1-12"; grayish to brown or reddish above, and with pale greenish or yellowish triangle on the nose and dark strip through eye; the outer hind toe is particularly broad. Habitats: rocky, cold, clear streams in humid forests or brushy areas. Sights: wandering in damp woods after rains; eyes seen shining in flashlight beam along edges of streams.

Chirping Frog

Leptodactylid or barking and chirping frogs, family Leptodactylidae. Either frog- or toadlike; toes beneath often have obvious tubercles; smooth and semitransparent eardrums; circular skin fold may appear on belly.

Greenhouse Frog

> Barking Frog (*Eleutherodactylus augusti*). 2-4"; as above; looks like toad, but has slender and unwebbed toes; also skin fold shows across back of head, and there is semicircular fold on belly; grayish purple on back, often with light, clouded appearance, and with dark blotches that have light borders. Habitats: rocky areas, caves, mines, juniper-mesquite areas, cacti, etc. Sights: peculiar stilted walk over rocky surface with heels high off ground; comes out on damp nights to hunt insects.

> Greenhouse Frog (*Eleutherodactylus ricordi*) and chirping frogs (*Syrrhopus* species). Tiny size, ⅝-1½"; light, chirping calls. Habitats: gardens, greenhouses; rocky cliffs, damp woods, palm groves, thickets of brush, etc. Sights: comes out at night to hunt insects; may be seen hopping, leaping, and sometimes running; very quick movements.

Ranges of
Chirping Frogs

Tree frogs and their allies, family Hylidae. The slim waists, quite long legs, and small size are distinctive; some have toe pads.

> Cricket Frogs (*Acris* species). ⅝-1⅜"; rather warty for frogs; green, brown, or gray in color, usually with V-shaped mark be-

Range of
Greenhouse Frog

Cricket Frog

Chorus Frog

Tree Frog

Gopher Frog

Range of Cricket Frogs

tween eyes, and a long dark stripe or stripes on hind side of thigh. Habitats: shallow pools, sloughs, bogs, swamps and their banks; rarely climb bushes. Sights: skittering over surface of water, taking long (3-4') and erratic jumps; singing in groups in the water at night.

Chorus Frogs (*Pseudacris* species). ⅔-1¾"; slim and small, without toe pads; variably colored brown, gray, green, etc., but with long dark stripes from head through eye and down back and sides; sometimes spotted stripes instead. Habitats: like grass in shallow water of ponds, lake edges, etc. but sometimes in deeper water; rarely climb, then only low; often on farms or in city parks. Sights: making rather short jumps on edge of water; swimming in semicircle back to land; hiding in grass.

Tree Frogs (*Hyla* species). ½-5½"; head relatively large, eyes large, waist narrow, toes with adhesive climbing pads; colors variable: brown, gray, green, often changing in some species to suit surroundings and give camouflage. Habitats: swamps, freshwater marshes, brushy areas, moist woodlands, ponds, gardens, cultivated fields, rocky canyons, etc. Sights: may come to lighted windows at night to seek insects; seen climbing bushes, small trees, rocks, using their adhesive pads to climb up steep surfaces; gather in ponds, usually where there are bushes or trees in water, to sing after rains; some fall from trees in pursuit of insects on damp nights.

True frogs, family Ranidae. Usually very long-legged with narrow waists, smooth skin and large webbed hind feet; usually great jumpers.

True Frogs (*Rana* species in part). 1⅜-8"; as above; body and legs usually long; variously colored greenish, grayish, and brownish, with red or yellow areas, and generally whitish to yellowish underneath. Habitats: usually found in or near water, but also in damp woods, brush, etc. Sights: making long jumps when disturbed; hunting insects by shooting out tongue or leaping on them; congregations sing in shallow water where water plants give cover.

Gopher and Crawfish Frogs (*Rana areolata*). 2½-4"; body short and plump; head comparatively large; legs short for frogs; usually yellowish gray or green with dark brown spots, or with darker gray-brown background. Habitats: near water, living in bank, burrows, or holes in ground of streamside woods by day. Sights: comes out of holes at night to hunt insects, etc., jumping rather shortly.

Range of Chorus Frogs

Ranges of Tree Frogs

Range of True Frogs

Range of Gopher and Crawfish Frogs

Exploring Life in the Desert

Deserts contain an amazing variety of life—reptiles and other animals, and hardy plants as well. A few days or even a week or so in the desert, particularly in late spring, and after a period of rains, can be an exciting time! At night the desert literally teems with life, and the noises to be heard and the sights to see are incredibly numerous. Despite the fact that such deserts have many poisonous creatures, including giant centipedes, scorpions, rattlesnakes, sidewinders, and (in Arizona) the Gila monster, all of these creatures are quite respectful of man and make way for him through knowledge that he is more dangerous than they are! Nighttime desert explorers need only to move slowly and watch where they are stepping in order to avoid all danger. A flashlight helps to spot these creatures and ankle-high boots give protection.

After a rare desert rain, if it is a fairly hard one, plant growth in the desert is often incredible (see Chapter 8 for pictures and descriptions of some of the desert plants). If you see a flower stem starting to rise out of a yucca plant, take a ruler or tape measure and measure it. Come back in a few hours and measure its astonishing growth. It often grows a foot or more in a day, fast enough to sit by it and even watch the movement of growth! Other plants like the cactus are swelling with water so that their spiny stems take on increasing fatness in a few hours. The Agave plant takes the water of a spring rain and rushes to completion its 18-foot stalk in three weeks or less, crowding the top with flowers that literally burst into bloom in one evening! Measure all these growths and see for yourself. Then watch and smell for the night-blooming cereus, a plant with flowers of giant white petals in an arrangement as much as fifteen inches in diameter! Each night in late spring a few blossoms open, the movement of the great blooms opening almost visible to the eye, and throwing out an almost overpowering fragrance that brings in the night-flying hawkmoths in excited crowds to probe the flowers for their nectar. Elf owls, poorwills, and other predators may come to prey on the nectar-seekers, so that a bewildering variety of life appears on the scene.

This life can be measured in various interesting and exciting ways. For example, check the number of visitors to one great blossom from the time it opens in the evening until you leave, so figuring the average number of visitors per hour, each bearing pollen from another flower and taking more pollen away for distribution to and fertilization of blossoms elsewhere.

Elf owls are found mainly where saguaro cacti grow to their immense heights of thirty feet or more, each elephant-shaped stalk bearing up to seven tons of stored water! In abandoned cactus woodpecker holes in these stalks, the tiny owl makes its home; and to its hole each owl returns several times a night with tiny prey. If one owl is seen departing and returning to one hole for several hours, it is certain that the bird is feeding its young a regular diet into the night. On a moonlit night, or using a red flashlight, which does not disturb the birds, you can watch its efforts for many hours, noting the number of trips it makes an hour and determining also the type of prey it brings. Thus a cricket or crickets show several long legs dangling out of the mouth; a scorpion has had its poisonous sting on the tip of the tail bitten off by a sharp beak, but the two pairs of pincers may dangle from the grasped and crushed body; a large centipede appears almost wormlike with its length except for its numerous legs; a tiny shrew shows its long-pointed nose; a mouse's thick body may be torn into three pieces or more before the elf owl can carry parts of a creature larger than itself, but a victim of its ferocious attack. Those who have watched this little red-eyed demon of the saguaro forest night swear that it is the most voracious hunter they have ever seen!

The gecko lizards, the Gila monster, the desert tortoise, and various snakes are the main reptiles of the desert night. Watching a gecko—small, smooth-skinned and weak in appearance, and not particularly fast as a runner—is to see a creature who has more enemies than can be counted on all ten fingers! A large scorpion is not adverse to attacking, stinging, and then holding it with large pincers, while the life juices are sucked out. An 8-inch centipede can easily destroy it also, while a giant tarantula will leap on it and plunge into it its poisonous fangs, to kill and suck the inner fluids. An elf owl is constantly ready to swoop on a gecko from out of the night sky; while a desert skunk, bobcat, or kit fox will leap to seize it. A grasshopper mouse—its head up to catch the scent of a gecko, while it gives a long tremulous squeak, like a wolf howling on the trail of a deer—comes racing after it, its short round tail quivering with blood hunger. How indeed can the gecko stay alive in the midst of so many fearsome night enemies? How can it be given time or peace to find and eat its own smaller insect prey?

By watching and following a gecko with a red flashlight, moving quietly so as not to disturb it, you may find the answer. Notice how its color design of black patches against brownish yellow gives it excellent camouflage amidst the dark rocks and yellowish plants of the desert floor. The instant it senses movement around it, it stops in its tracks and freezes so that it truly appears like a part of its surroundings. Another thing a gecko does if it thinks it is being attacked, is to lift its tail up and wave it about stiffly, giving the impression that it is a scorpion and so frightening away an inexperienced fox or wildcat. On top of this, if actually attacked, the tail breaks off and goes wriggling off by itself, thus often attracting the predator while the real lizard quickly hides and then stays still! A new tail is grown back in a few days. A fourth aid against attack may be the squeaking cry the gecko often makes in the night. This startling sound may frighten some would-be attackers, or it may be used as a signal of danger from one gecko to another. Watch and determine the answer yourself.

Safe watching can be done in the desert night, by preparing a four-legged platform about 2-3 feet high on which a chair is fastened, the legs of the platform being each inserted in a large can filled with oil. By sitting quietly and comfortably on this chair with a pair of binoculars and a red-shielded flashlight, you can observe most of the movements of life in the surroundings. None of the poisonous creatures of the night can possibly reach you on this chair.

As snakes, a desert tortoise, or a Gila monster come into sight, time their movement with a watch as they pass by marked and measured posts or stakes you have placed around your watching place. If a tortoise, for example, takes one minute to cover twenty feet between two stakes, then it is traveling at a speed of slightly less than one-quarter of a mile an hour. It is obvious from the slowness of the tortoise that it has means other than speed to avoid enemies, and you can see its defense in operation when it becomes disturbed and pulls legs, tail, and head inside its carapace in such a way that it would be hard to attack it inside the hard shell defense. Watch its eating habits also and determine how long it takes to eat its plant food. Is it as deliberate about that as about its movements? If the time of observation is the dusk just before dawn, the desert tortoise may decide to dig itself a hole in some soft ground where it can get away from the heat of the coming day and also avoid daylight enemies, such as man. How long does it take to dig each foot into the ground? Time it and see.

The snakes that come and go in the night may be quite different in their actions. The great western diamondback rattlesnake comes slithering through the dark with all the strength and forwardness of a tiger in its native jungle, knowing that almost all things that live fear its passage. The bright-banded king snakes have a similar bravado, since they are renowned killers of rattlesnakes and other reptiles. The little dark-colored night snakes are liable to slip along in the shadows and among the rocks, keeping their bodies mostly hidden. The sand snakes and shovel-nosed snakes come out of the loose sand to hunt for crickets and other insects, but dive and swim beneath their protecting sand cover whenever danger threatens.

The sidewinder rattlesnake is the most distinctive of all. He does not need to hide beneath the sand because of his poisonous bite, but he does need to travel rapidly over the shifting surface after his mice and gecko prey. So he lifts and slides his body in repeated J's, like an expert ballet dancer, the loops hissing over the sand with a curious repeated pattern of sounds almost like a whispering machine.

Besides checking the speeds of these creatures, try mapping on a piece of paper the directions of their movements—whether straight, curved, zigzag, or otherwise—and so gain an inkling of their character. Thus the mysteries of the desert night gradually unfold.

CHAPTER FIVE

Exploring Fish
Night Life Styles

Probably MORE FRESHWATER fish are active by day than by night simply because more need light to catch their active insect prey, particularly the kind that alight on the surface of the water or come near it. However, many fish and other aquatic forms of life are more active during the nighttime than during the daylight because they can escape their daytime enemies at night. Some fish thus avoid many of the larger herons and the kingfisher, with his sudden plunging into the waters with powerful bill, or the similar plunging of pelicans. Insects also are often more active at night in the waters for the same reason. On moonlit nights fish come in numbers to the surface to catch insects falling or flying there, and there is much activity under the water too as the moon filters its beams beneath the surface film. And there are some fish, like catfish, who use feelers and other sensory organs beside eyes to get about in dark or muddy waters.

Spawning Habits and Hatching Methods

There are many ways to distinguish the various families and species of fish, and some of the differences between kinds of fish are very interesting. For instance, fish differ widely in their spawning habits and the ways in which they hatch and protect their young. Some salmon and the steelhead trout move upstream night and day at certain times of the year to their spawning beds, often traveling in large numbers seriously intent on covering distance. They swim quite fast. Pacific lamprey eels, on the other hand, although they swim upstream from the ocean to lay their eggs in nests among the rocks of swift streams, travel more slowly and sluggishly and not in such large numbers.

Fish have developed two ways of laying and protecting eggs from destruction; both methods can be observed in clear streams on moonlit nights during the egg-laying seasons, perticularly as some fish like to be surrounded by darkness when laying their eggs because of the greater protection afforded. Many fish lay the eggs in great masses along the rocky or pebbly bottoms of streams or shallow lakes and ponds, and leave them to make their way in life without further help. Salmon, bass, etc. use this way. The idea is that out of these many thousands of eggs at least some will not be eaten and will live to be adults.

The other method is to build a nest, as do the sunfishes, catfishes, sticklebacks, and some of the minnows. This nest may be little more than a hollow area or depression dug in the sand or gravel in which the female deposits eggs, or it can be a partially woven nest of water plants. In some cases the male guards the eggs from enemies until they hatch, but in other cases the male or both male and female guard the young fish after they are hatched until they get big enough to take care of themselves. In these cases the female has to lay far fewer eggs than in the random spawning method first described. In between these two methods is that of some trout who lay large clusters of eggs, but at least partially cover them with sand and gravel to protect them before they leave them.

A third way is not to lay eggs at all, but for the female to hatch her eggs inside her body, later giving birth to live young. This gives still more protection to the young so that a fewer number need to be produced to make sure some will survive. Watch for all these ways under the waters when you spy on fish at night.

Variation in Size

Fish also differ greatly in size, and for this reason the Guide to Sights of Fish and Their Relatives later in this chapter gives the length of all fish described therein. Our freshwater fish range in size all the way from an inch-long killifish up to a 10-foot-long white sturgeon of the Pacific Northwest, which may weigh more than half a ton. But it is hard to know when to say a fish has reached adulthood, as many of them continue growing all their lives. Thus a 10-inch trout may be an adult, but it will grow on for several years up to as much as 20 inches or more, if it is lucky to stay clear of hooks and other enemies. Generally speaking, lake fish grow bigger than stream fish. A lake trout, for example, may grow over 3 feet long and pass 80 pounds in weight, while 2 feet in length and 30 pounds is enormous for a stream trout.

Differences in Food-finding Methods

Differences between fish in the methods by which they find food and attack their prey are many and fascinating, and will well repay the study of beginning nighttime explorers. Trout are noted for their ability to leap after flies, whereas suckers are rather sluggishly moving fish that find their food by scraping and sucking it off rocks on the bottom of streams. Catfish, on the other hand, are slow movers also but catch their food in rather muddy waters of large rivers by feeling for it with the long feelers or barbels attached to their heads. These various habits are noted in this chapter and can help not only in identifying these fish but also in learning more about the interesting life pattern of each.

Identifying Fish in Relation to the Continental Divide

In studying the descriptions of fish in the Guide to Sights of Fish and Their Relatives further on in this chapter, note whether the fish are found east or west of the Continental Divide. This is the great dividing line; fish west of this line should not be confused with those east of it. Usually, they are quite different in species and even in genera. Some kinds of fish are found in just one kind of water drainage, as in the Sacramento and San Joaquin water basin of California, or the fish of the streams in the Great Basin in Nevada and Utah that enter into inland lakes or sink into the sands. In the East, on the other hand, the water systems of the Great Lakes and St. Lawrence River are closely connected with the vast Mississippi-Missouri River systems, so that over great areas, very similar fish are found. Even the rivers that empty into the Arctic have some connections with the St. Lawrence system, so similar fish exist from the Arctic Ocean to the Gulf of Mexico.

Preference of Fish for Different Types of Water

Remember that some fish, such as the salmon and trout, like cold water best and are rarely if ever found in warm-water lakes or streams. Black bass and sunfish, on the other hand, like warm waters, even up to 75 degrees and higher, and so are found in lowland lakes and rivers where these temperatures are encountered. Some fish—such as the lake trout, the Great Lakes whitefish, and the lake herrings or ciscos—insist on living only in the still waters of lakes. Various darters, the brook trout, and some other fish, on the contrary, prefer only small, swift streams where there is lots of oxygen in the frequently disturbed water.

Identifying Fish by Fin Arrangements

One of the best ways to tell fish apart is by body shape and arrangement of fins. The fish silhouettes in the Guide to Sights of Fish and Their Relatives below will help the reader to identify fish by this means. The Guide identifies some of the very spectacular or common species of fish, plus other groups of fish, each of which have rather specific shapes and fin arrangements. The accompanying generalized picture of a fish shows the names of the different fins. It will be useful for you

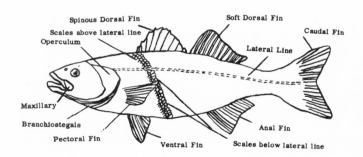

Parts of a Fish

to study it, as fish are often identified by the position, size, and shape of these fins. Thus the bass has two large dorsal fins, closely connected; while a trout has a spiny dorsal fin and a soft dorsal fin reduced to a small fleshy protuberance. On the other hand the freshwater burbot is instantly recognized by its very long soft dorsal and anal fins far back on the body. It thus pays to watch for and study these fins very carefully when observing fish.

GUIDE TO SIGHTS OF
FISH AND THEIR RELATIVES

(Note: As fish are not too readily detected in the water at night, only a few of the very common kinds or genera are listed here, and are pictured by shape or silhouette rather than in detail. Dimensions refer to length.)

Lamprey family. Eellike with large sucker-disc for a mouth.

Lampreys (Lampreta and Entosphenus genera). 7-25"; the long narrow bodies have fins only near hind end. Sights: seen wriggling way slowly upstream in migrations.

Lamprey

Sturgeon family. Shovel-shaped nose, rows of armor plates on top and sides; suckerlike mouth with row of sensory barbels in front of it.

Sturgeons *(Scaphirhynchus* and *Acipenser* species). 2-12' long; as above. Sights: usually seen swimming in large lakes, or scraping along bottom with mouth in clear waters.

Sturgeon

Herring family. Distinctive sawtooth edge to front of belly; bluish back and silvery sides.

Shads or Alewives and River Herrings *(Dorosoma* and *Alosa* species). 5-30"; as above. Sights: swimming in rivers, hunting small animal life.

River Herring or Alewife

Salmon family. Has soft fleshy structure called the adipose fin near the tail on back, and the pelvic (or central belly) fin has an axillary process (or long, pointed structure) at its base.

Trout

Salmon and Trout *(Salmo* species, *Salvelinus* and *Oncorhynchus* species). 1-5'; weight: up to 100 lbs.; generally swift-swimming, cold-water fish with bodies dark and spotted above, silvery below and on lower sides. Sights: leaping from water at night after insects, especially on moonlit nights.

Salmon

Whitefish, Lake Herrings, and Chubs *(Coregonus* species). 6-16"; more silvery general color or olivaceous than trout and salmon and with comparatively few spots. Sights: leaping from water after insects.

**Whitefish of
Coregonus Species**

Smelt family. Also have adipose fins, but generally more slender and silvery bodies than the salmon and trout.

Eulachon Smelt

American Smelt

Smelt *(Eulachon, Osmerus, Hypomesus,* and *Spirinchus* species). Up to 12″; as above, but some with dark backs. Sights: most often seen swimming in large spawning runs in rivers in the spring, but some in lakes; river seems to be running with silver when they spawn, often at night.

Pike family. Characterized by duck bill-like jaws and a large soft dorsal fin, far back near the tail.

Pike

Pikes and Pickerels *(Esox* species). Up to 60″ and over 75 lbs.; as above; generally found in large rivers and lakes. Sights: savagely chasing other fish.

Sucker family. Have typical small suckerlike mouths without teeth but thick lips that can be extended; the dorsal fin has usually more than 10 rays; they are very bony, but edible.

Sucker

Suckers *(Cycleptus, Ictiobus, Carpiodes, Erimyzon,* etc. species). 8″ to more than 3½′; as above. Sights: lazily swimming along bottom sucking up edible debris; swarming up creeks in spring in spectacular numbers.

Minnow family, Cyprinidae. Though most minnows are small, as their name implies, a few reach 3-5′. They do not have suckerlike mouths, though some look as if they do; most have less than 10 rays in the dorsal fin, except a few, such as goldfish and carp (both introduced from Europe), and most have no spines. It seems best for the purpose of this book to simply picture here a few typical common minnows to give an idea as to shape and size.

Carp

Redbelly Dace

European Carp *(Cyprinnus* species). Up to 32″.
Dace *(Rhinichthys* and *Semotilus* species). Up to 5″.
Suckermouth Minnows *(Pehacobius).* Up to 4″.
Redbelly Daces *(Chrosomus* species). Up to 5″.
Squawfishes *(Ptychocheilus* species). Up to 2′.
Gila Daces, Shiners, and Chubs *(Gila* species). 3-15″.
Typical Minnows *(Pimephales).* 2-4″.
Typical Chubs *(Hybopsis* species). 2-10″.
Shiners *(Notropsis).* 2-7″.

Young Pickerel

Squawfish

Gila Thicktail Chub

Typical Minnow

Shiner

Catfish family. Have no scales; sharp spines on back and sides, flat and broad heads and long barbels around the mouth.

Catfish *(Ictalurus, Pylodictus, Noturus* species). 1-5'. Weight: up to about 100 lbs. Spines give irritating and painful wounds, but poisonously so only in *Noturus*. Sights: seeking food on river-bottoms at night, using barbels as feelers.

Catfish

Freshwater eel family. Very long and slender, with dorsal fin very long from tail tip to past midbody, and continuing under body.

American Eel *(Anguilla rostrata)*. Up to 3'; brownish yellow and as above. Sights: great numbers of females sometimes seen swimming up rivers in fall; sometimes wriggle about on nearby land on damp nights for hours.

Stickleback family. Slender-bodied and with row of sharp dorsal spines in front of short dorsal fin; pelvic fins are strong spines.

Sticklebacks *(Pungitius, Eucalia, Gasterosteus,* and *Apeltes* species). 2-4"; as above. Sights: Pugnaciously guarding nests from other fish and animals; hunting small creatures of shallow waters.

Stickleback

Killifish and topminnow families. Small fish of both salt and fresh water, some deep in body and some quite slender, but all have mouths that point upward for use in surface feeding and somewhat pushing outward lower jaws; all have incomplete or only partially developed lateral lines.

Common Killifish *(Fundulus* species). 2-6". Mosquitofish *(Gambusia* species). 1-1½". Sights: eating at surface of water.

Bass and sunfish families. Rather high spinous dorsal fin, usually completely separate from soft dorsal fin; typical bass shape (for bass); typical thick sunfish shape.

The Basses *(Roccus* and *Micropterus)*. 15-20"; dorsal fins mostly separate.
The Sunfishes *(Lepomis, Mesognistius,* etc.). 3-12"; dorsal fins joined and front spinous dorsal fin generally much shorter.
The Crappies *(Pomoxis)*. 8-12"; strongly mottled dark green or black on silvery. Sights: all these fish are surface feeders, seen gulping insects on surface at night.

Bass

Sunfish

American Eel

Common Killifish

Mosquitofish

Crappie

Yellow Perch

Pike Perch

Darter

Sculpin

Perch family. Perches, pike perches, and darters. Distinctive division of dorsal fins into two well-separated and about equal-sized portions.

Yellow Perch *(Perca)*. 2-12"; body yellowish with strongly marked dark greenish bands; pinkish belly. Sights: commonly seen breaking surface with snout to seize low-flying insects.

Pike Perch or Sauger *(Stizostedion* species). 12-36"; has many long sharp teeth; body only weakly cross-barred. Feeds mainly on other fish, which it chases.

Darters *(Percina, Etheostoma, Ammocrypta* species). 2-6"; usually with slender bodies; and strongly marked dots or blotches along lateral line. Sights: dart swiftly from side to side in shallow water to escape capture.

Sculpin family, Cottidae. Characterized by large flat heads with eyes close together and perched high on head; pectoral fins quite large and fancy-looking; no scales on body, but often have tiny prickles.

Sculpins *(Cottus* and *Myoxocephalus* species). 3-7"; generally beautifully mottled above. Sights: come out from under their rock hiding places at night to feed on small life of pond and stream, which they attack by rapid movements.

How Fish Survive Oxygen Depletion

Fish are hard to find both day and night in wintertime in northern areas where ice gathers on the lakes and ponds. In these areas the oxygen in the water is greatly depleted, especially at shallow depths. This is because the green plants that produce oxygen and keep adding it to the water are quiescent in winter. Man contributes to the oxygen depletion problem when he pollutes streams and other bodies of water. Pollution kills a lot of fish by taking away their oxygen supply.

During the winter fish, water insects, and other water creatures usually go into dormancy in shallow northern ponds, scarcely staying alive and so using very little oxygen, in the hope that they will be able to exist until the ice breaks and plants start growing again. This is one reason why such a pond in the first warm days of spring shows such an explosion of life. An observer who came with an underwater flashlight and studied the life in the pond just before the warmth began and then came again a few days later would be amazed at the way life had burgeoned all over the place. He would see schools of fish in excited chases after water insects and worms where before there was practically nothing.

Exploring Underwater with Lights

Lights are essential in studying underwater life at night. It is possible to go to the neighborhood electric shop and have them produce a long extension cord with a completely waterproof light globe attached to it. This cord and light must be prepared by experts because electricity can be very dangerous if it comes in contact with water and can give a severe shock. An underwater flashlight is a good deal safer, but the light from this is not as powerful or useful for exploring below the surface.

In either case, the object is to get this light under the surface of a stream, pond, or lake at night and observe the actions of the underwater denizens. If a cord is used, fasten it to a wooden spear which can be driven into the bottom of the pond so the light is held low and it is not necessary to touch the cord once it is in the water. A white light is useful for attracting a large number of fish and insects. They come swimming towards this strange and fascinating center of light for much the same reason that moths and other insects are attracted to a light hung in a park, backyard, or wilderness area. Into the beam of white light, if it is held steady so it does not frighten, come swimming curious fish, gyrating or diving insects, and possibly even crayfish or turtles, sticking their heads out of holes among the rocks or plants of the bottom. All, at first, are curious. Then gradually the larger, faster, or more ferocious creatures begin to use the light as an aid to catch the smaller creatures that are brought to the light, and a deadly game of life and death begins. Small newcomers, dazzled by the light, may be swiftly gobbled up, but others, learning from experience, or using long-held instincts of flight or dodging, avoid their enemies. In this little underwater world is concentrated the essence of the story of life and its struggle to exist by both eating and escaping being eaten.

If, on the other hand, the object is to watch the life under the night waters as it goes on in more ordinary circumstances, then cover the white light with a red plastic film. Many creatures which are active mainly at night will pay no attention to this red light because it is literally not visible to their eyes and will go about their natural habits undisturbed. The discreet observer, on the other hand, by sitting still by the water's edge so as not to disturb them, can see them plainly. Here among the fronds and stems of the water plants is seen the voracious water tiger, the larva of the dragonflies, sneaking about, ready to throw its extensible jaws out in a flash of steellike pincers to seize a passing worm, insect, or small fish. Or a lurking trout may suddenly close in on the water tiger and swallow it with one gulp, showing another link in the chain of underwater life.

Add to the pleasure and clearness of what can be seen by making a waterproof box with a glass bottom and fastening this so the glass is a few inches below the surface and the light is caused to shine beneath this window into the depths. Then sit or lie in comfort and see with much greater clarity all the fascinating adventures of living things that are happening below the surface, for the surface film and the ripples no longer interfere with your vision. You may even discover things about the habits of fish that naturalists who have watched and studied them only by day are not aware of. Good hunting!

CHAPTER SIX

Meeting Insects and Their Relatives at Night

THE SWARMING OF insects at night can often be astounding. In countrysides where they have not been too much killed off by poison sprays, myriads come to the lights on summer nights, often filling every cubic foot of space with dancing wings. Other dozens and hundreds come to sweet baits of molasses and fermented beer painted on trees and on strips of cloth hung to branches. Through the grass stems and weeds of the miniature jungles found in meadows and gardens, fields and parks, other thousands are found hiding, hunting, and fleeing. Armies and columns of ants march on the surface of the ground, while dead wood may be filled with the cities of the termites. In deserts, during the time of great heat, far more insects appear at night than during the daytime.

The section that follows on life in the miniature jungle of a weed patch will provide some idea of what might be seen and the tremendous interest of exploring such places at night.

Law of the Jungle Among the Insects

Down among the grass stems in a meadow or deep in the small jungle made by a cluster of weeds in a vacant lot, life picks up in activity during the nighttime, especially if the summer day has been a hot one. A red-covered flashlight is a key to open up this little world and see the activities and adventures of many living things.

The ants are particularly active in this small jungle. They crawl over the ground as individual scouts and hunters, or move in columns after food or to attack other ants. On the bushes there are lines of them going upward to reach the leaves and milk their ant cows, the plant lice or aphids, each of which has a tube or tubes extending from its rear end from which the ants can cause a sweet drop of liquid to come by stroking the rear of the green body.

The ants are constantly on the hunt for new sources of food and, if one of them finds such, it quickly carries the news to its fellows. For example, an ant may come suddenly on one of the most interesting sights of the weed jungle, the hatching of a whole batch of praying mantis babies from their eggs which have been protected through winter and storms in a stout covering of hardened foam excreted around them by their long-dead mother. Now the brown plasticlike covering is breaking loose into cracks from which pour the baby mantids, coming out into the darkness which would usually protect them from the sight of enemies such as birds. But tonight a single ant has spotted them and away he races down the stem and over the ground to tell his people, knowing instinctively that these creatures, who will later be so ferocious and deadly and heavily protected by hard armor, are now soft and weak at the moment of birth.

The tiny mantids begin to climb all over the stems and leaves of their plant home, their triangular heads twisting around in the notable and unique mantis way. You see their front legs forming naturally the praying gesture of their kind, uplifted and curved in apparent supplication, but sure forerunners of the time when the sharp leg prongs will close like steel traps over many an insect and even frog or young mouse. Some in their dawning hunger even attack a brother or a sister and kill and eat it on the spot. But a greater danger is coming: the ants streaming over the ground and then up the main plant stem in two ruthless columns. The tiny mantids, soft now of outer skeleton, whose chitin is only beginning to harden, are seized quickly by the greedy ants and crushed and chewed in snapping jaws. One by one they are carried back to the ant nest as fair game for the victors. Fortunately, a few escape by running over the leaves and down other stems, seeking with instinctive wisdom for whatever hiding places they can find.

All about in the darkness other tragedies, explorations and adventures are happening. In the night-blooming primroses, and similar night-opening flowers, hawkmoths are coming for their meals; and sometimes waiting for them in the flowers are secret demons of the night who crouch in ambush—bloodthirsty crab spiders, hidden deep in the petals, or even more ferocious assassin or ambush bugs, waiting to spear the moths with long sharp proboscises that act like front-ended stings. On the floor of the little jungle, ground beetles are running about hunting for fallen caterpillars or worms they can eat; and sexton beetles come flying when they smell the body of a recently dead mouse or bird that they can set to work burying for their future young to feed upon.

Crouch low, watch carefully, flash the red light to all sides, and you will be amazed at what is happening all about.

Because so many kinds of insects are active at night, people who are just beginning to learn about the world of the outdoors after dark may wonder just where to start in knowing about nocturnal insects. Such beginners will have to limit the scope of their investigations at the outset. This book helps them do so by listing some of the most common and outstanding kinds of insects and their relatives that can be seen at night. The books about insects listed under the Suggested References at the back of this book provide more detail on the lives and identification of these creatures.

The species and groups of insects listed and illustrated in this chapter are those that are most likely to be seen. They are arranged under the particular habitats or special circumstances in which they are likely to be encountered at night. Study the shapes shown in the illustrations to familiarize yourself with what may be seen. But remember this is only the beginning of one of the most complex and varied of all subjects to be studied at night.

COMMON INSECTS AND RELATIVES ATTRACTED TO LIGHTS

Geometrid Moth

Arctiid Moth

Moths
Geometrid moths
Arctiid or tiger moths
Noctuid or owlet moths
The notodontids or prominent moths
Pyralid moths

Wasps and Ants
The horntails
The sawflies
Ichneumon wasps
The ants

Horntail Larva

Orders of Homoptera and Hemiptera (plant bugs and relatives)
The lantern flies
The assassin bugs
The water boatmen
The giant water bugs

Order of Neuroptera (the nerve-winged insects)
The mantispids
The lacewings
The ant lions

The Beetles
The ground beetles
The rove beetles
The soldier beetles

Horntail Adult

Ichneumon Wasp

Lacewing

Ant Lion

Ground Beetle

Rove Beetle

Black Ant

Assassin Bug

Water Boatman

Giant Water Bug

Mantispid

The checkered beetles
The click beetles*
Scarab and June beetles
Long-horned beetles

Order of Orthoptera (grasshoppers, mantids, cockroaches, etc.)

The cockroaches
The praying mantis
The crickets

Order of Ephemeroptera (the mayflies of many families)

Order of Plecoptera (the stoneflies of several families)

Relatives of the Insects

Spiders
Tarantulas
Wolf spiders
Others
Solphugids

COMMON INSECTS ATTRACTED TO SWEET FORMULAE†

Moths

Catocala moths‡
Tarache moths

Checkered Beetle

Click Beetle

Scarab Beetle

June Beetle

Long-horned Beetle

Cockroach

Praying Mantis

---

* Click beetles make a clicking noise and turn over by a snapping movement when put on their backs.

† Paint the sweet mixture of fermented beer and molasses, well mixed, on leaves, bark, cloths hung from limbs, etc. during the day. At night, come back with a flashlight, or, better, a head lamp, and find what treasures have come to your lures. Move quietly and slowly, for the creatures of the night may flee at a sound, and you will miss many beauties. A moist, warm night with a hint of storm in the air is a specially good time to go, for many a living thing rises in joy to the soft, damp warmth of the night air.

‡ These moths have beautiful bright banded underwings.

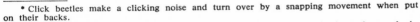

Catocala Moth

Solphugid

Tarantula

Stonefly

Wolf Spider

Field Cricket

Mayfly Nymph

Mayfly Adult

Geometrid Moth

Tarantula

Solphugid

Scorpion

Wolf Spider

Black Ant

Centipede

Sphinx Moth

Pyralid moths
Geometrid moths
Mamestra moths

Ants

INSECTS COMING TO SPECIAL FLOWERS*

Sphinx Moths†

Flower Beetles

Ground Beetle

LARGE INSECTS AND OTHER INVERTEBRATES CRAWLING OVER GROUND, PARTICULARLY IN DESERT AREAS

Beetles

    The ground beetles
    Scarab beetles
    Carrion beetles‡

Ants and Wasps

    The ants
    Mutillids or velvet ants§

Scarab Beetle

Spiders and Similar Invertebrates

    Tarantulas
    Wolf spiders
    Solphugids
    Scorpions
    Giant centipedes

Carrion Beetle

Black Ant

Mutillid

---

    * The special flowers include the evening primroses, honeysuckles, and night-blooming cereus.

    † These moths come often at high speed or may hover in front of flowers, putting deep into the corolla their long tongues to suck up the nectar.

    ‡ These brightly colored red and black beetles work through the night to drag the bodies of small dead birds and mammals to soft ground and then bury them with herculean effort, which is most exciting to watch.

    § The wingless females are very hairy and are especially found around the nests of bees and wasps, which they parasitize.

Mayfly Nymph

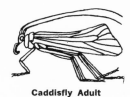

Caddisfly Adult

Predaceous
Water Beetle

Water Strider

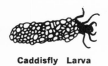

Caddisfly Larva

Whirligig Beetle

Water Scavenger
Beetle

Water Boatman

INSECTS AND THEIR RELATIVES
SWIMMING IN POOLS OR WALKING ON BOTTOM

Insects

    Water striders
    Water boatmen
    Giant water bugs
    Backswimmers
    Water scorpions
    Mayfly larvae
    Stonefly larvae
    Caddisfly larvae
    Predaceous water beetles
    Whirligig beetles
    Water scavenger beetles

Giant Water Bug

Spiders

    Fisher spiders*

Crustaceans

    Crayfish†

Backswimmer

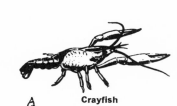

Crayfish

Shellfish and Snails

    Freshwater snails
    Freshwater clams

Freshwater Snail   Freshwater Clam

Water Scorpion

INSECTS AND THEIR RELATIVES
CRAWLING ON TREE TRUNKS AND STEEP ROCK SURFACES

Insects

    Noctuid moths
    Carpenter ants

Carpenter Ant

Stonefly Larva

    * These spiders run about on the water surface or climb down plants below the surface; they build rafts in the water from which to watch while looking for insect prey.

    † Crayfish can be seen crawling in and out of holes among rocks or water plants and waving long antennae and large pincers.

Snowy Tree Cricket

Katydids
Snowy tree crickets

Snails

Land snails

Firefly Beetle

INSECTS COMING TO SCENT OF FEMALE FROM LONG DISTANCE

Giant Silk Moth Males* (Saturnids)

INSECTS THAT FLASH LIGHTS IN DARKNESS

Firefly Beetles†
Lantern Flies
Glowworms‡
Click Beetles§

Glowworm

Click Beetle

Katydid

Giant Silk Moth Male

---

* Males have very large feathery antennae; females much smaller antennae. If a freshly metamorphosed female is placed in a jar with a mesh top near an open window on a warm spring or summer night, males will come to her from miles away.

† Males flash lights while flying above females on ground, who reply by also flashing lights.

‡ Glowworms are the larvae of fireflies; they make little dots of light in the darkness.

§ These beetles may have glowing green headlights and a red tail light.

CHAPTER SEVEN

Knowing Life
by the Waterside

THERE IS NO better place to study and observe many of the forms of life which this book has covered in the preceding chapters than the shore of a body of water—whether the body be an ocean or a pond. Numerous animals and birds already described come to the seashore at night to gather food, particularly the raccoon, mink, gray fox, weasel, otter, bobcat, coyote, skunk, muskrat, short-eared owl, great horned owl, most of the shore birds, the marsh birds, the gopher snake, and the king snakes. The shore birds follow the retreating waves to dig quickly into the sands for worms, small crabs, ghost shrimps, sand fleas, and other hidden denizens. The marsh birds come where a creek or river mixes its water with the sea, and find their food among the fringing plants by probing with their long bills in the mud and among the plant debris. Much shore life is more active by night than by day. During the hours of darkness shores may swarm with moving creatures.

Aside from the visitors mentioned above, the seashore has considerable life of its own. The life of the seashore varies from birds and mammals, which are high on the scale of evolution, clear down to the simplest kinds of life like the hydroids and the sponges.

Zones of Life on the Seashore

All life on the shore is keyed to the great rhythm of the tides. The best time to come to the shore, of course, is when there is a good low tide, for then a lot of unique life is exposed to the beams of flashlights. On the other hand, however, a lot of creatures on the shore are most active when the tide is coming in, for that is when the sea brings them food. Microscopic creatures and plants in the water, called plankton by the scientists, are brought in by the waves and are eaten by larger creatures, which, in turn, may be eaten by still larger creatures. Life and death have the same rhythmic movement as the tides and the waves.

So we can start at the bottom of a low tide and see life active where the little waves are tumbling into the lowest tide pools, then retreat slowly as the tide comes in, avoiding being caught in the big waves, and watching life being revived into frenzied activity as the water comes first to the low tide zones, then the middle tide zones, and finally the high tide zones. In the low tide zone life is adapted to being constantly wet, for the sea makes its presence felt there most of the day; but in the higher tide zones life has to deal with longer and longer periods of being away from the touch of the sea, so that the outer skin or armor becomes tougher, and many kinds of life must hide in deep crevices or under seaweed to preserve the precious dampness.

How Seashore Organisms Protect Themselves

Each bit of life is unique and has special ways of making its living. Watch particularly how the organisms of the seashore feed and protect themselves against enemies, for these activities will help in both identification and understanding of these creatures and their adventurous life. Thus, mussels and barnacles generally like to gather on the rocks in large social groups, often very close together to give them protection from both wave action and the attacks of predators who find them harder to get at when they are jammed close together. The mussels, in particular, form with their close-packed bodies ideal protective jungles in which slim-bodied creatures—like rock fleas, mussel or nereid worms, and porcelain crabs—move like the animals of a real jungle, some as hunters and others as hunted. In the lower pools tube worms form similar close-packed colonies on the rocks for the same kind of protection, while free-swimming animals like the blennies (tiny fishes) and the shrimps depend on swift sudden movements and camouflaging colors that look like the water plants to escape their enemies, such as larger fish and crabs.

The following section on the mysteries of the seashore will give a glimpse of this life.

Mysteries of the Seashore Night

Few places are so mysterious and ghostly at night as an uninhabited seashore. Somehow the very noise of the sea itself is strangely different, the sounds of the water both louder and more intriguing. The little waves seem to laugh, and the deep sigh of the sea resembles the long breathing of a great sea monster. Then threateningly the roar of the cresting, turning waves comes ominously out of the darkness, and the foam comes riding in on the waves with a mysterious light that fills the beaches with glimmering lines.

Of more importance to the watcher, however, is the vast amount of life that comes out at night on the beach, in the tide pools and along the edge of the sea. This is because the night gives protection, particularly from the gulls and terns, who cover the beaches in hordes during the day, picking up and eating every living thing they can find. So out from under the hiding fronds of the seaweed and out from the cracks and caves of the tide pools come myriad forms of life rarely seen by daylight. Armies of crabs now cover the rocks, while sand fleas swarm and leap on the sandy beaches. Every pool is ablaze with vivid life, as tiny tide pool sculpins or rock pool johnnies, garibaldis, blennies, and other small fish dart through the dark waters, sometimes leaving trails of light behind them if the sea is filled with the fire of phosphorescent microscopic animals and plants. Shrimps also stab through the waters, flung about by the snap of their tails that often saves them in a tenth of a second from the clash of eager jaws or pincers. Great cancer crabs and the powerful-pincered kelp crabs are the tigers of these tide pools, lurking like monsters to spring from their hiding places in caves or under the fronds of the seaweeds on unsuspecting lesser creatures. All about is the adventure of life, the constant escape or capture of the hunted and hunter.

Perhaps the greatest mysteries are discovered in the sea caves and the deeper pools at low tide. It is best to come on those low tide nights when the sea is relatively calm, as a slippery rock at night on the edge of the sea is no place to be when a great wave comes. And if there is any hint of danger at all, go in a party in which all are tied together with rope, each person fastened to the rope with an unslipping bowline hitch. When the tide begins to rise, come back to the safety of the shore. Don't stay out of eagerness to explore and get into trouble with those advancing waves.

In a deep pool is likely to be seen one of the most marvelously designed of all seashore hunters, the great sunflower starfish, with eight to twelve to twenty-four arms or more, each moving the animal along over the rocky surface with a constant flowing, shimmering motion that makes the whole creature more the dream of a fantastic imagination than a flesh-and-blood living thing. Soon a sunstar may be found swarming over a large mussel or other seashell and use the powerful suction of its myriad tube-feet to get an unbreakable hold on the shell. Applying remorseless pressure, the great starfish, sometimes as much as eighteen inches in diameter, gradually forces the shells of the bivalve apart, everts its own stomach to put it inside the open shell, and digests the helpless shellfish. So do all starfish eat their shellfish prey, but, in the sunstar, the motion and digestion is easier to see and faster and more spectacular.

On the walls of a low-tide cave every square inch is covered with forms of rooted animal life, many looking far more like gorgeous flowers than true animals. Lovely Metridium anemones wave their white tentacles in the green waters like a thousand swaying dancers. Tube worms reach out even more delicate and feathery tentacles, often brilliant red in color, from the ends of their calcareous protective tubes. The tiny flowerlike heads of pink-hearted hydroids wave snakelike in the gently moving waters along the lower cave walls, while next to them the stiff spines of a green sea urchin move spasmodically to ward off the slow but remorseless attack of a purple sea star. Nearby a four-inch-long plumed sea slug, its back fantastically covered with a forest of waving orange papillae above a grayish orange body, moves over the red- and green-marked surface of a crumb-of-bread sponge to attack a larger forest of stinging anemones.

The incredible beauty of these underwater and tide pool gardens, in which plant seems to turn into animal, and animal into plant, often in the twinkling of an eye, and the colors of leaf- and flowerlike structures and waving tentacles and papillae change from moment to moment with the changing of the waters and the shifting of the flashlight's beam, makes such an evening spent on the edge of the sea more filled with wonder than a visit to fairyland or the castle of a sea king.

GUIDE TO SIGHTS OF
COMMONLY SEEN SEASHORE CREATURES

(Note: to avoid confusing the novice, this chapter lists and describes only the most common kinds and groups of seashore creatures, with silhouettes to show their general appearance, and explanations of their special habitat niches on the shore. Notice that the descriptions of seashore life not only tie in the various forms of life with the tide zones but also tell whether they are associated with rocky shores or sandy shores and pinpoint the special parts of those shores—tide pools, exposed rocks, seaweeds, and so on—in or near which the organisms can be found. Dimensions given in each description refer to length, unless otherwise specified.)*

Fish.

Fish of the tide pools or the sands.

Blenny

Blennies *(Anoplorchus, Epigeichthys, Xiphister,* and *Gibbonsia* species).* 3-8"; very slender and eellike, slippery in the hand, usually of dark colors to blend with dark seaweed. Sights: darting about in tide pools at night after small creatures.

Clingfish

Clingfish *(Gobiesox meandrica).* 2-4"; sucking mouth, dark colors. Sights: slithering over the rocks, or clinging like a leach by suction to rock surface when waves break. West Coast.

Sandgoby

Sandgoby *(Clevelandia ios).* 2-3"; pale sandy color. Sights: comes out of hole in sand to hunt small creatures when tide is in. West Coast.

Garibaldi *(Hypsypops rubicunda).* 3-5"; bright gold color. Sights: hunting through tide pools at night. West Coast.

Rock Pool Johnny

Rock Pool Johnny or Tide Pool Sculpin *(Oligocottus maculosus).* 3-5"; reddish brown and with pretty patterns; big ugly head, pointed body; big pectoral fins. Sights: darting out from under seaweed. West Coast.

Grunion *(Leuresthese tenuis).* About 3-8"; grayish silver in color. Sights: comes to spawn on the coast of southern California on the second, third, and fourth nights after the full moon during the period from March to June; literally thousands swarm up on the edges of the waves in the sand to lay eggs, female and male together as a pair. The female digs a shallow hole in the sand to lay the eggs, while the male fertilizes them.

Grunting Fish or Midshipman *(Porichthys notatus).* 4-6"; dark with light buttons on sides. Sights: comes out at night from under rocks to hunt small life; makes grunting noise when disturbed. At night shows row of buttonlike lights on sides. Northwest coast.

Sea Horse

Sea Horse *(Hippocampus* species).* Up to 6"; horselike head; body held vertically when swimming. Sights: swimming slowly through seaweed jungles near shore. Atlantic coast.

Atlantic Croaker *(Micropogon undulatus).* Up to 14"; silvery gray with thin black vertical bars on sides. Sights: feeding on crustaceans and seashells in shallow water near shore, and in seaweed. Makes croaking noise, heard when ear is held close to water surface.

Marine Tide Pool Sculpins *(Clinocottus analis)*. Up to 12"; large ugly head, very spiny and usually large fins. Sights: moving slowly about tide pools near bottom at night.

Marine Tide Pool Sculpin

Puffers or Swellfish *(Sphaieroides maculatus)*. Up to 2'; swell to large size by sucking in water when alarmed. Sights: feeding on crustaceans in shallow water. Atlantic coast.

Burrfish or Spiny Boxfish *(Diodon holacanthus)*. Up to 11"; covered with small spines; can also swell sides out for protection. Sights: feeding on small creatures in low tide pools.

Crustaceans.

Lobsters, crabs, shrimps, etc.

Lobster *(Homarus americanus)*. Up to 2'; dark greenish and spotted above; first pair of claws very large for defense or seizing prey. Sights: seen on summer nights digging holes in substratum or emerging cautiously from holes or under rocks to scavenge for dead or living creatures. Atlantic coast.

Shrimps. Up to 2"; very long second pair of antennae; front pincers much smaller than lobster's.

Lobster

Common Shrimps *(Crago species)*. Length to about 2"; generally greenish gray in color; body more compressed than in crab or lobster. Sights: shooting through water of tide pools by explosive movement of tail.

Pistol (Snapping) Shrimps *(Crangon species)*. Up to about 2"; noted for peculiar construction of large claw enabling shrimp to make loud snapping noise to frighten enemies. Sights: darting under plant cover in tide pool when light is shown on them.

Common Shrimp

Broken-Back and Transparent Shrimps *(Spirontocaris species)*. ½-2"; back bent as if broken, but actually used to hurl shrimp backward through water; smaller sizes of higher tide pools are almost completely transparent. Sights: interesting to see internal organs working through transparent walls of body; or to see shrimp snap its back to thrust itself backward into shelter.

Pistol Shrimp

Crabs, generally with powerful pincers and flat wide armored bodies (except hermit crab).

Hermit Crab *(Pagurus species)*. Up to 3"; characteristic partly curled soft body which is usually inserted into empty snail shell for protection, while armored front part of body fills entrance. Sights: mainly seen tumbling about tide pools scavenging for food, or fighting over food and shells with each other in rather comical ways; swarm over beach at night looking for food to scavenge.

Broken-Back Shrimp

Edible Crabs *(Cancer species)*. Up to 8"; characteristic shape (as shown) like a wide-based fan; generally reddish color. Sights: active at night in low tide pools, scavenging and hunting small living things with large pincers.

Hermit Crab

Edible Crab

Blue Crab

Ghost Crab

Fiddler Crab

Shield-backed Kelp Crab

Shore Crab

Horseshoe Crab

Beach Hopper

Skeleton Shrimp

Blue Crab *(Callinectes sapidus)*. Up to 7″ broad; shell oval; very large claws. Sights: running sideways; swimming with extreme speed for a crab; males carrying females on back and protecting them from other males. Atlantic Coast.

Ghost or Sand Crab *(Ocypode albicans)*. Up to 3″; sandy color, square shell. Sights: digging burrows in sand; running down to waves to get gills wet in the night; feeding on dead and live food brought in by sea tides.

Fiddler Crabs *(Uca* species). Width: ½-1⅔″; single very large pincer in male. Sights: male courting female in elaborate waving of large claw, luring her into hole; large claw used to threaten both enemies and other males. West coast.

Shield-backed Kelp Crab *(Pugettia producta)*. Up to 4″; shiny olive green color; U-shaped shell; very agile with powerful pincers, so dangerous to handle. Sights: found hiding in seaweed, but comes out at night to hunt for small life in tide pool. West Coast.

Shore Crabs *(Pachygrapsus* and *Himigrapsus)*. Up to 2″ wide; shell rather square-shaped, reddish or greenish and often with stripes. Sights: very active at night scavenging for food and eating algae on rocks, putting food in mouth rapidly with large front pincers in rather comical way; raise pincers threateningly if attacked. West Coast.

Horseshoe Crab *(Limulus* species). Up to 1½′; horseshoe shape with long tail spine. Sights: females drag males up on to beaches on dark spring nights to lay eggs and have them fertilized. Atlantic coast.

Beach hoppers and skeleton shrimps, as described below.

Beach Hoppers   *(Orchestia, Hyale, Talorchestia,* etc.). ¼-1″; all have typically curved body and use strong hind legs (two pairs usually) for leaping; body the color of the sand or of sea plants, mussel colonies, etc. where they live. Sights: particularly active at night, when they hop about the beach in thousands or burrow into piles of debris, scavenging for food brought by the sea.

Skeleton Shrimps *(Caprella* species). ¾-1¼″; color of seaweed in which they live; body elongated and skeletonlike. Sights: seen by flashlight waving and bowing their bodies on fronds of seaweed as they catch tiny creatures brought in by the sea.

Mollusca.

Seashells, snails, etc.

Sea snails, order of Gastropoda. Either bear a coiled shell on their back or are sluglike creatures with the shell small and hidden.

Nudibranches. Colorful sea-sluglike creatures, with gills forming tentaclelike structures on back.

Aeolid Nudibranch

Aeolid Nudibranches or Sea Slugs *(Hopkinsia, Flabillena, Her-missenda, Dendronotus, Aeolis, Coryphella,* etc.). 1-5″; the plumelike gills extend most of the body length; may be bright yellow, blue, green, and other colors. Sights: crawl over similar colored sea sponges or sea plants.

**Dorid Nudibranch of
Genus Glossodoris**

Dorid Nudibranches or Sea Slugs *(Glossodoris, Rostanga, Lamelli-doris, Archidoris,* etc.). ½-4″; plumelike gills form a cluster on the rear back; also very colorful yellow, green, bluish, etc. in many combinations. Sights: as above.

**Dorid Nudibranch of
Genus Archidoris**

Sea hares and bubble shells, etc. Some appear sluglike, but some have bubblelike shells. None with waving gill-like structures on back as in nudibranches; bodies are large. All these are West Coast creatures.

Limpet

Bubble-Shell Snails *(Bulla, Haminoea,* etc.). 1-2″; the bubble-like shell resting on top of the large body, or partly surrounded by it, is typical. Sights: feeding on seaweed in tide pools.

Sea Hares *(Navanax* and *Aplysia* species). Up to 1′; colorful purplish brown *(Aplysia),* or with bright blue lines and yellow dots and dashes *(Navanax)*; eyes at tips of earlike tentacles at front of head. Sights: *Navanax* may be seen gobbling up the smaller bubble-shell snails; *Aplysia* feeds ravenously on various seaweeds.

Periwinkle

Other gastropod sea snail shells. Most of these are found in low tide zones or below where they are not often seen; mentioned here are a few of the upper tide zone most likely to be seen at night.

Limpets *(Acmaea* species). Up to 1½″; oval, flat pyramidal shells, not showing coils. Sights: move about slowly over rocks or eelgrass eating tiny algae; have rasplike tongues or radulae.

Periwinkles *(Littorina* species). Up to 1″; gray to brownish or greenish; shell with medium-shaped spire (as shown). Sights: crawling slowly about on rocks, often above high-tide mark, when tide is in, eating algae and other plant life.

Turban Shell

Turban Shells *(Tegula* species). Up to 1½″; usually brown or black in color, sometimes with lighter markings; shape turban-like (as shown). Sights: moving about over rocks in sheltered high-tide pools or gathering in great clusters; sometimes seen floating down a thread of mucus to bottom of tide pool. West Coast and Florida.

**Moon Shell and
Sand Collar**

Moon Shells or Sand Collar Shells *(Polinices* species). Up to 4″ tall; large whorls below, small whorls like cap at top. Sights:

Scallop

Rosy Jackknife Clam

Shell and Siphon of Bent-nosed Clam

Basket Cockle

burrow in sand, also come out to make sand collars with mucus in which they embed their eggs.

Bivalves or clamlike seashells, order of Pelycypoda. Two shells attached together by hinge.

Scallops *(Pecten* species). Up to 6″ wide; distinguished by earlike projections on each side of the shell and row of eyes where shell opens. Sights: seen swimming in shallow water by clapping two halves of the shell together, forcing out jets of water that hurl the creature backward at high speed; repeated, it shoots off at another angle. Rows of eyes along the shell edge warn it of danger. When hunting, it moves slowly forward. Most intelligent of the clamlike shellfish.

Clams *(Mya, Macoma, Solen, Siliqua,* etc. species). Up to 10″; various shapes. Sights: usually bore in mud of substratum, extending long "feet" to surface, where food is gathered; sometimes shoot water high in air when exposed at low tide.

Cockleshells *(Cardium* species). Up to 5″ wide; typical heavily flanged and grooved cockleshells, fanlike or heartlike in shape. Sights: often seen traveling at surprising speed over rocks on large feet.

Echinoderms.

Common Pacific Starfish

Webbed Sea Star

Sunflower Star

Starfish, brittle stars, sea urchins, sand dollars, sea cucumbers.

Starfish. Starfish arms are rather broad and used to move animal slowly over rock, while tube-feet on underside act as suction cups to hold it to the rock against pull of surf.

Common Pacific Starfish *(Pisaster* species). Up to 18″ in diameter; tough leathery skin of orange, purple, or yellow coloring, etc. High- to low-tide zones. Sights: feeding on mussels and other seashells by forcing open shells to pry at insides with everted stomach. Pacific shores.

Webbed Sea Star or Bat Star *(Patiria miniata).* Up to 6″ in diameter; flat and wide appearance is distinctive. Sights: in middle and low-tide pools on rock walls, moving slowly, or trying to force open shells of shellfish to eat insides. Pacific shores.

Sunflower Star *(Pycnopodia helianthoides).* Up to 24″ in diameter, with as many as twenty-four arms; beautiful orange color. Low-tide zone and below. Sights: many arms move rhythmically as it moves over bottom of tide pool greedily gobbling up small creatures. Pacific shores.

Sun Star *(Solaster* and *Crossaster* species). Up to 14″ in diameter, with 8-15 arms; many colors, often very beautiful. Middle to low-tide zones and below. Sights: similar to those listed under Sunflower Star above.

Sun Star

Common Atlantic Starfish *(Asterias* species). 4-12″ in diameter; many colors; body leathery; 5-8 arms; rows of tube-feet on underside of each arm. High- to low-tide zones. Sights: attacking oyster by humping body and applying pressure with arms and tube-feet from opposite sides, pulling oyster shells apart until it can evert stomach and pour digestive fluids in on body of oyster till latter is digested.

Sea urchins. Basket-shaped shell is covered with movable spines.

Common Atlantic Starfish

Sea Urchins *(Strongylocentrotus, Arbacia* species). 2-8″ in diameter; purple, red, green, etc.; tube-feet in separate clusters; pedicillerae, or tiny teeth, hidden among spines. Sights: poke at sea urchin with sharp point and spines converge to fight off point; poke at sea urchin with dull point and spines open to enable pedicillerae to bite at point. They feed on seaweeds and defend themselves against attacking starfish with pedicillerae and poison.

Sea cucumbers. Have long soft or leathery worm-like bodies with ring of short-branched tentacles around the mouth; tube-feet adapted for creeping on one side of body only.

Sea Urchin

Sea Cucumbers *(Cucumaria, Stichopus,* etc.). Up to 18″; usually black, brown, red, or green. Sights: body buried in substratum, but tentacles around mouth waving about like flower petals in strong breeze; may explode insides out of body if badly disturbed; body generally flaccid when calm, but becomes stiff and turgid when attacked, and much shorter in length; water that is pumped into the anus may be pumped out again in a great spurt or stream.

Sea Cucumber

Annelida.

Segmented worms. Most live hidden in burrows, but a few more often seen ones are mentioned here.

Clam, mussel, and sand worms *(Nereis* species). Up to 18″; usually greenish, but also various other colors; many-segmented, each segment with bristlelike appendages; have sharp horny jaws at tips of extensible proboscises. Sights: voraciously hunting and eating small creatures, including other worms of tide pools; swim through pools and nearby ocean waters at night with sinuously graceful movements, especially during breeding season.

Mussel Worm

Tube Worms *(Sabella, Serpula, Hydroides,* etc.). Up to 16″; usually in calcareous or parchmentlike tubes which adhere to the rocks or substratum. Sights: tentaclelike or filamentlike gill brachia are put out from head end of tube into the water to collect tiny ocean life brought by the tide. Since these are often of varying and even brilliant colors of red, orange, blue, etc., when they are seen waving in the water at night in the light of a flashlight, they make a striking sight.

Tube Worm

Coelenterates.

Jellyfish, hydroids, sea pens, sea anemones, etc.

Jellyfish of the Genus Aurelia

Jellyfish *(Aurelia, Cyanea* species). Semitransparent jellylike bodies, some up to 10" broad, but most much smaller; many with ethereal colors of blue, green, pink, etc., generally with hanging tentacles. Sights: wierdly pulsating as they swim slowly through the dark waters of large tide pools or in the open sea.

Sea Anemone

Sea Anemones *(Corynactis, Bunodactis, Anthopleura, Metridium,* etc. species). Up to 20" in diameter; many colors, but most often white or bluish green; with numerous tentacles surrounding mouth hole; tough leathery body. Sights: waving tentacles as tide comes in, or closing rapidly around a small animal the tentacles have captured; throwing out shell of crab or seashell from mouth after the soft parts have been digested.

Wildlife Dramas by a Pond in the Woods

Like the seashore, the margins of a freshwater pond abound with an amazing variety of creatures, many of which are more active by night than by day. A full-moon night by a real wilderness pond in the summertime is a place of magic and action. Such a place usually has many habitats: underwater life; surface-water life; pond-side plant jungles of cattail, water lily, and bur reed; pond-bank life under the willows, cottonwoods, and bushes that live near water; often a meadow that borders the pond; broad-leaf woods of maple, oak, and hickory; often a coniferous forest that brings its dark and somber green ranks down to the shore; also bogs and marshy lands at the upper end of the pond. Here is a good way to watch the action:

Select a place to sit on a folding camp chair or a piece of foam rubber partly leaning against a tree. It should be where several different habitats can be seen at once. Have mosquito repellent if necessary, but remember that mosquitoes usually go away later in the evening. Netting around your face and over a hat may also be used. Be sure to rub both body and clothes with a strong-smelling plant that hides the human smell. Assume a comfortable position and be very quiet and without movement for hours if necessary!

The moon turns the waters of the pond to silver, shimmering beauty, but all may be quiet for a while, for your approach will have been seen and heard by many creatures. Yet, if you sit perfectly still, soon the animals of the area will begin to act normally. A frog will jump suddenly into the water nearby and then swim out toward a lily pad. A deep croaking from one or more bullfrogs will begin at the end of the pond. A bat will come swishing through the air over the pond and disappear as quickly as it has come. Then, magically, other bats will appear as if the first one was a scout sent to find out if enemies are lurking by. Their wings will swish and their jaws will click as they snap up moths or other insects, but you will very likely not hear the high-pitched squeaking by which they judge through the echoes of bat sonar where each insect can be found flying. But you will see them sweep with uncanny accuracy to meet in midair their swift-flying prey, and by watching intently, you will see some moths and other insects escape them by dizzily whirling or dropping down into the bushes or into the leaves of the trees before those deadly jaws can snap.

If the evening is still early or the moon very bright, nighthawks will also zoom over the pond, depending on high speed and wide-open jaws, instead of swift dodging, to seize insects in midflight. Hairs on the sides of the birds' large mouths act as traps, driving the flying creatures into the open maw, which may even look like an inviting hiding place to a frightened moth, recently escaped from a bat!

Now a dark V of ripples forms on the waters and a dark head forms the point of the V as a muskrat comes swimming towards the water plants. He will dive down, seeking not only for bulbs to eat, but also for succulent freshwater clams, mussels, and snails he finds on the bottom and brings up to place on a rock to dry until they open their shells with frantic despair at finding themselves out of water. Or he may use his large front incisor teeth to chomp through the stem of a large bulrush or cattail that he then carries away to his stick and mud house on top of the water in a secret, water-plant-hidden part of the pond.

After the muskrat has departed towards his home, a darker and more slender form may appear swimming where only recently the muskrat left his silver trail of water bubbles. It is probably a mink whose nose has picked up the scent of his natural prey, and is off on a search that may end in a kill. Careful listening may even pick up the sounds of battle when the muskrat squeals with both anger and fear, and a clicking and grinding rend the air as the sharp canine teeth of the mink clash with the even larger and more powerful front incisor teeth of the muskrat in the narrow tunnel leading to the muskrat nest. If the muskrat is old, large, and wise in the ways of fighting, the mink may even come off second best and depart from those narrow quarters with a savage growl of frustration.

Suddenly a swarm of mayflies come dancing over the water from some place among the plant stems where they have recently changed from underwater nymphs to winged adults. They are performing their mating dance, a dance that can lead only to death since the adults are not equipped to feed themselves, but only to produce fertilized eggs that are laid on the surface or dropped onto submerged stones. Their dance causes a frenzy to appear on the surface of the waters, for the fish, both great and small, rise to this natural bait from their lurking depths in the waters; many leap into the air after mayflies, splashing the water into waves and ripples and flashing their scales in the moonlight. But so many myriads of eggs are laid that some hundreds will eventually hatch into underwater nymphs that hunt for tinier creatures among the muddy waters around the plant stems and themselves are hunted by numerous larger animals, from dragonfly nymphs to crayfish.

A long enough wait in complete silence may be rewarded by the coming of a pair of otters, driven almost mad with delight by the moonlight, into the pool. Unlike the businesslike mink, whose mind is usually on food, the otters couldn't care less on such a fantastic night. They are out for fun, leaping and frolicking through the water like a couple of boys in a magic playground, seeming even to leap up to try to grab the moon, so high do they throw their sinuous brown bodies. Racing joyously, they go to the side of the pond where a steep clay bank soon produces ludicrous antics as they climb up to slide down it, sometimes right-side up and sometimes on their backs, twisting and turning, so eager to get back to the gay task of sliding that they literally split the pond surface with their foaming tracks through the waters to hit the shore and gallop up again to the top of the slide.

Suddenly the male gives a sharp whistle and the two vanish with a splash into the dark waters. Maybe you moved for just an instant and he saw you, or maybe the train of shadows coming out of the woods is what scared him. But it is only a family of raccoons, mama, dad and the four kids, all of them scamps at playing tricks on each other and as curious about any new creature or thing they can find on the pond edge as any boy that ever lived! The incredibly active, facile hands of the young are soon playing in the water and feeling underwater plants and stones in careful imitation of their elders, who are seeking crayfish and frogs. One raccoon kid lets out a yelp because the pincers of a crayfish suddenly got him instead of the other way around! But mother comes to the rescue with one sharp wrench of her paws that breaks the crayfish's back, and then the whole family is greedily tearing it apart to get at the delicious meat inside the shell. Soon another kid whines as he jerks out a hand to show some dripping blood, but it is only where a large brown leech had sunk its circular row of small teeth into the skin to get blood from a vein and then been torn loose by the kid's jerk. So do the young of all species learn from experiences that there are some dangers and unpleasant things in life that need to be avoided if possible. Father raccoon then gives a lesson in frog catching, showing how to cautiously approach a frog sitting on a leaf until near enough to make a sudden grab. If he misses, he dives into the water, whirling those grasping hands down to seize the frog even when it is swimming. But many a big and wise one gets away!

On a trail by the pond and leading into the fir woods the raccoon family suddenly meets a skunk family coming the other way, the long black and white tails waving over the heads of the skunks, both big and little. There is a churring growl from the mother raccoon, but it is a warning rather than a challenge, and she shows the necessity of being polite to those black and white furred creatures by moving to the side of the trail to let the new family pass. An inquisitive raccoon youngster noses forward to investigate, but is knocked back quickly by his elder into his proper place, just as the mother skunk stamps her feet in warning! Raccoons are not afraid of skunks, but they don't want to argue with them either!

In the moonlight in the meadow a flurry of movement shows a group of cotton-tail rabbits suddenly getting up the courage to do their full-moon night dance. But even in the madness of the moon, they try to stay close to the forest edge and its hiding shadows and bushes. Over and over each other in great hops they jump, one diving under just as the other goes up and over—all in perfect sychronization of movement. As the mad movements increase, and they become more forgetful of danger, it seems as if one circle of leaping figures is going opposite to another circle within the first circle. But, before you can decide whether this is really so, the rabbits' challenge to the fearsome night is answered. A great black shadow sweeps low from the surrounding woods, coming with absolute noiselessness. There is a shrill scream and the rabbits disappear as if their dance had never been, for the great night tiger of the air, the horned owl, has taken his nightly tribute, the fearsome claws and powerful beak almost instantly killing the struggling rabbit who had forgotten to watch for danger. Yet the rabbits had to make this challenge and this sacrifice. Death came, but so also was the dark and life itself made more exciting and keen for those who escaped.

The whole pond and all its surroundings is a great circle of life and death, of eat and be eaten, of hunt and be hunted. And this is a part of the magic of the moonlight, for every rustle or squeak of a mouse, every call of a bird, every soft sound of a furred foot moving down a trail, every whisper of wings and leap of a fish from the waters is tied in with the quest for food and life. It is a drama to be watched for hours without fear of satiety as long as one's eyes can stay open and alert.

CHAPTER EIGHT

Examining Plant Life After Dark

EVERY PLANT HAS unique qualities that make it different from every other living thing, so the nighttime explorer can find something of interest in each plant he encounters. But some plants seem to have special values that make them stand out, and a few of these are mentioned and described here to increase awareness of the more interesting qualities some plants give to the night. Of course, other plants—like thorn bushes, cacti, nettles, poison oak, poison ivy, and poison sumac—should be avoided both day and night. If you feel the sting of nettles or the sharp points of thorns, instinct, if not good sense, will cause you to back away. Just be careful not to back away into another bunch of the same plants!

Certain very common plants set the dominant tone of various habitats or plant communities, and there are books (listed in the Suggested References at the end of this book) that describe and picture not only the plants but also the common animals likely to be found in these communities. Thus various live oaks are dominant trees in the oak woodlands of the Far West, while balsam firs and black spruces dominate the great Canadian coniferous forests that are also found in some of our northern states from Minnesota to Maine.

How Plants Move in the Wind

In looking at plants in the night, observe their shapes and how they respond to the wind. The shapes as well as reactions to wind, spell, to the watchful observer, the qualities of plants and help him appreciate them better. Knowing them gives a sense of having old friends about and may give a flavor to the night that will live longer in memory. Thus oaks give a feeling of staunchness and solidity. They move in the wind under protest, fighting it as does a powerful boxer responding to the blows of a lighter opponent. But the willow is far different, standing slender and gracefully with easy bows to the lightest breeze, its whole nature that of willingness to give before blows.

Nighttime Personalities of Trees and Forests

In the nighttime plants, especially trees, take on qualities that they may not have or have to a lesser degree during the day. The gloom of the deep redwood forest, for example, or that of the Sitka spruce forest on the northwest coast, is so all-pervading that, without the use of a flashlight or other light, the observer seems as if lost in a fathomless pit. And he senses, even if he sees only the vaguest outlines, that he is surrounded by massive giants whose tops disappear into an unkown world far above him. In an eastern deciduous forest, on the other hand, a person feels surrounded by waves of flickering light, particularly if there is a wind blowing and some moonlight, for each of the light-colored leaves is catching some light and flashing it to other parts of the forest to produce an effect like a gigantic kaleidoscope. The dimness of the light at night adds a feeling of mystery and magic that is absent during the day, and, if one were suddenly to find himself surrounded by the little folk of the woods, it would not seem unnatural. In the desert the giant saguaros, often over thirty feet high, look like several-armed giants, and their massive size so contrasts with the smaller desert plants as to make the observer seem to be in a land of Lilliputians invaded by Gullivers. Indeed, the night there is alive with many little folk, particularly kangaroo rats and pocket mice.

Plants as Shelter for Animals and Birds

Look at every plant not just as a plant, but as a possible home and shelter for animals and birds. Hollow trees and woodpecker holes often shelter various creatures. If you tap on the wood outside of such a place, murmuring or growling protests may come from within, and a flashlight turned inside could show interesting things. Under the leaves of other trees and in the hollows where branches come together birds and other creatures often hide or sleep. In the West, arboreal salamanders may hide in such places, and in the Southeast are different forms of tree frogs that climb in the trees and some that even get so anxious chasing insects up among the leaves that they miss jumps and fall on your head like rain!

In the pages that follow, the leaves of particularly interesting plants are shown to help in identification. The books listed under Suggested References will help you identify many more.

IDENTIFICATION GUIDE TO PLANTS

(Note: Other plants are described in Chapters 9 and 10.)

Molds and fungi.

Slime Molds (*Lycogala, Stemonitis, Criraria,* etc. species). Some of these appear as thin slimy sheets literally crawling over wet logs, and may show phosphorescent lights in the dark.

Slime Mold of Genus
Lycogala

Slime Mold of Genus
Stemonitis

Bald Cypress

Mushrooms (various genera). Some of these, such as glistening coprinus *(Coprinus micaceus)*, gleam brightly in moonlight, forming fairy rings; others may literally glow with phosphorescent lights on rainy nights.

Gymnosperms or coniferous trees.

Bald Cypress *(Taxodium distichum)*. Distinctive trunk supported by great buttresses, common in southeastern swamps. Sights: Buttressed trunks and drooping branches give mysterious or ghostly feeling to southeastern swamps at night; hollows in trunks often form shelter or dens for wild creatures, particularly raccoons and owls.

Big Tree and Redwoods *(Sequoiadendron giganteum* and *Sequoia sempervirens)*. These enormous trees have reddish, very thick bark and small cones. Sights: massive trunks and great height give feeling of timelessness and power to western forests at night.

Junipers and Red Cedar *(Juniperus* species). Sprawling shrubs or many-branched trees with one single trunk; bark shreddy, usually many dead brittle branches. Sights: distinctive angular shape and many branches, silhouetted against sky on western Great Plains and mesas or high foothills of the mountains, give shelter to many creatures. The eastern red cedar is not so conspicuous, as it is usually mixed with other trees.

**Awllike Leaves of
Red Cedar**

Monocotyledonous or parallel-veined plants.

Palmetto *(Sabal* species). Height up to 85′; distinguished by bracketlike circular rows of old leaf bases on upper part of stem; fan-shaped leaves as much as 8′ long with numerous long narrow

**Cone, Leaves, and
Silhouette of Big Tree**

**Juniper Leaves
and Fruits**

**Scalelike Leaves and
Fruits of Red Cedar**

**Palmetto Leaf and
Fruit Cluster**

Greenbrier

Leaves and Acorns of Northern Red Oak

Leaves and Acorn of Bur Oak

Leaves and Acorns of Turkey Oak

Joshua Tree

leaflets. Sights: the distinctive massed thickets of these palms form wonderful shelter and hiding places for wild creatures. Southeast.

Greenbrier *(Smilax* species). Woody vine, green-barked, winding and climbing on trees or other bushes; covered with sharp thorns; oval to almost round leaves with parallel veins. Sights: a part of dense thickets, keeping men out with the thorns and so protecting much wildlife.

Joshua Tree and Other Yuccas of the Southwest *(Yucca* species). These odd-shaped plants are usually covered with sharp spiny leaves; usually with large white flowers. Sights: distinctive silhouettes at night in desert country, short grass plains, or chaparral; give excellent shelter and food to many animals and birds.

Dicotyledonous or net-veined plants.

Trembling Aspen

Spiny Hopsage

Trembling Aspen and Other Poplar-Type Trees *(Populus* species). Up to 120′ high; leaves characteristically almost round, but with sharp points and generally light green or yellowish green in color. Sights: leaves tremble in every breeze, especially those of trembling aspen, so appear to ripple and shimmer constantly in the moonlight.

Oaks *(Quercus* species). Up to 150′ high; rather broadly shaped trees with unusually thick central trunk; unique acorn-shaped nut. Sights: leaves, light green below and dark green above, show contrasting patterns of light and dark and impart a sense of power and strength; shelter for much animal life, especially in hollow trees.

Spiny Hopsage *(Grayia spinosa)*. Shrub of up to 3′; has sharp spiny-tipped twigs; a desert plant with a unique bright green appearance, different from grayish or grayish green of surrounding plants. Sights: seems to glow with a soft green light in moonlight or dusk; attractive to deer, mountain sheep, and other wild animals, which may be seen grazing on it.

Sycamore

Hawthorn

Blackberry

Acacia

Screw Bean

Night-flowering Catchfly *(Silene noctiflora)*. Up to 3', with sticky coarse stems; lower leaves up to 5" long and narrow at base. Sights: Small white to pinkish flowers bloom at night, attracting hawkmoths in numbers.

Sycamore or Plane Tree *(Platanus occidentalis)*. Up to 175' tall; heavily mottled bark distinctively scales off in sheets. Sights: along streams the massive mottled trunks stand out like ghostly figures in the moonlight; the great tree may house many an interesting animal and bird.

Hawthorns *(Crataegus* species). Up to 20' high; has very spiny and crooked but slender branches; flowers like small apple blossoms, but generally with disagreeable odor; found along hedgerows, abandoned farms, etc.; of special interest because they are evolutionarily diversifying at a tremendous rate, saving cut-over and burnt land, and forming admirable shelter and food for many wild animals and birds. Sights: thickets of hawthorns are particularly favored by cottontails, which run in and out of their depths.

Honey Mesquite

Blackberries, Raspberries, etc. *(Rubus* species). Well-thorned shrubs and vines, often growing in dense thickets, with delicious red and black berries. Sights: attracting many animals and birds to their ripening fruits, but repelling humans by their sharp thorns; many animals such as rabbits and wood rats hide in their thickets.

Acacia, Screw Bean, and Mesquite *(Acacia* and *Prosopis* species). Usually very sharp-clawed plants of desert areas, with long narrow bean pods that rattle in the wind. Sights: great clusters of these grayish ghostlike plants enliven the desert washes and canyons of the Southwest, where they shelter and feed innumerable small creatures and birds.

Dogwood *(Cornus* species). Small tree or large shrub, up to 40'; what look like large white or pinkish flowers, as much as 5" across, are actually large bracts surrounding tiny white flowers in the center. Sights: these large white or pinkish "flowers" seem to glow in the dark during flowering in April through June.

Silhouette, Leaves, and Fruit of Dogwood

Honeysuckle

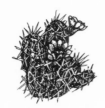

Cactus of Genus
Echinocereus

Cactus of Genus
Opuntia

Jumping Cholla

Flowers that open their petals mainly at night.

Evening Primrose

Night-blooming Cereus

Cactus of Genus
Echinocactus

Honeysuckles *(Lonicera* species). Shrubs or vines up to 10' tall, with freely branching light-colored twigs, attractive flowers borne in pairs; usually reddish fruit. Sights: attractive to hawkmoths and white-tailed deer at night.

Evening Primroses *(Oenothera* species). Shrubs to 6' or taller, with cluster of leaves up to 7" long at base; 4-petaled flowers open at dusk. Sights: opening of flowers as dark comes is remarkable sight, as plant appears almost animallike. Attracts hawkmoths.

Night-blooming Cereus *(Hylocereus* or *Selnicereus* species). Trailing or climbing vine as much as 45' long; has enormous flowers up to 15" wide which bloom at night. Found escaped from nurseries in warmer states. Sights: attractive to sphinx moths and other insects at night; flowers seem to glow in dark, especially on moonlit nights.

Cactus Flowers *(Echinocactus, Echinocereus, Opuntia,* etc. species). Numerous sharp spines over thick leathery stems; beautiful flowers after rains; stems hold water during dry seasons. Sights: spines shine in moonlight, giving warning to stay away. Jumping cholla *(Opuntia bigelovii),* actually seems to jump at you when touched and can give serious wound.

Night-flowering Catchfly (see above, under Dicotyledonous or net-veined plants.)

CHAPTER NINE

Listening In on Nature's Nighttime Sounds

THE ADVENTURER IN the world of nature after dark must first of all be at home in the dark and unhampered by unreasonable fears. Becoming familiar with the sounds of the dark—the sounds of the landscape or seascape itself and of the creatures that inhabit it—helps in this acclimatization process. But it is not enough to feel at home outdoors at night. One must also have some idea of what he is watching or listening for.

Obviously, it is easier to hear than it is to see at night. This chapter describes many of the sounds which birds, insects, mammals, reptiles, and amphibians are most likely to make at night—not only voice sounds but also the noises they make with their bodies. Of course, these sounds cannot be completely understood in isolation but must be studied in conjunction with the movements and appearance of animals, described in the preceding chapters, and the odors emitted by animals and plants described in the following chapter.

The Sound of Wind and Waves at Night

Among the commonest noises heard at night are those made by the wind as it caresses or buffets the land and its vegetation or the waters of lakes, rivers, and oceans. Listening to these sounds deepens one's awareness of the mystery and poetry of nature.

The sound of the wind alone on the land is different from the sound of wind mixed with the sound of waves on a great lake or the sea. At night all these sounds seem to be sharper than during the daytime, probably because sounds carry farther at night and also because, in wild places, man, with all his noises of machines and his shouting is not so likely to be around. The observer who comes silently into such places and listens becomes gradually a part of the wind and the wave. He listens and hears a thousand voices in a thousand different ways and gradually begins to understand the meaning of earth and sky, for the wind and the waves sing about them.

On the prairies of our middle states the wind blows through the high grasses with a song altogether different from the one it sings in the forest. In the high grasses of the prairies it calls with a keening, up-and-down song caused by the whipping down of the long grass stems before gusts and their rising again when short calmer periods follow. But in the short grasses of the Great Plains the wind has more of a deep crooning that makes a listener feel, as may be really true, that it is coming out of a thousand miles of empty space and has poured over the land unendingly, building a song of distance and the very limits of earth and sky. Sometimes a trace of sadness can be heard in that song, as if the wind were remembering the days when there were no fences, the buffalo herds wandered over the plains in clouds of dark bodies, and the Indians crept to the ridge tops to spy on them before making the long charge of the hunt.

The sound of wind in the brushlands of America is also different from the sound in a forest because there is less give to the bushes. Their stiff branches are close to the ground; and their leaves too are generally smaller, stiffer, and firmer than the leaves of forest trees, so that the sound of the wind going through them is firmer and steadier, a solid humming almost like an electric motor in a generating plant. In the broad-leaved woodlands there is a soughing or a laughing of the low wind and a screaming of the high wind, tearing at the big leaves as if they were ragged flags. And when the wind slowly dies, it makes a long, softening, rustling sound as if a million leaves were dying with it.

In the great forests of pines and firs there is a seething sound of a strong wind that only needles can help produce, for somehow the needles, especially the long ones, turn and twist and twirl in an endless turbine motion through the air that may at last create a roaring sound in a very high wind as if the forest were shouting to the sky. But in a low wind the pine and fir forests sigh and even moan a little like a deep-breathing giant struggling with a feeling of pain. And it is most wonderful to stand in such a forest in what starts as an absolute calm and then hear the trees beginning to sigh and moan far away until a gust comes traveling across the treetops and the voice of the sky seems to come down in a long shiver of deep breathing sound and the listener senses as never before that the whole great forest is alive.

The sound of the wind on the beaches is different from the sound of the wind and the waves out in the open sea, for the rocks and cliffs catch and enlarge the echo of the roar from the sea and may increase the sound of howling from the wind, while the coarse grass on the sand dunes whirs as the wind goes through it. The medley of sounds on a beach in a storm grows more and more jumbled and fantastic. The crash and roar of the waves building up to those towering giants that can sweep completely over a fifty-foot rock create a terrifying din, competing with the shriek of the wind. The wind blows the scraggling pines on the cliffs over into upside down L's with the tip of each tree streaming at right angles landward and their every needle screaming in protest.

How different it is when there is only a whisper of a breeze blowing over the sea or large lake, and the waves are only small rolls in a glassy expanse where the moonlight paints a shimmering trail as far as the eye can see. Now is heard the true sighing and breathing of the sea and sky—a murmur from the sea, and a whisper from the sky, yet with the repeated rhythm of a great heart beating far away across the edge of earth. The little waves that come up on the beach hiss so softly as they are sucked down into the sands that they are barely audible, but listen carefully. They are talking of the great whales of distant seas halfway around the world and about forgotten isles so hidden that men see them once in a century or perhaps not at all.

Communications Systems of Animals

The noises that birds, mammals, reptiles, and amphibians make are only part of their means of communicating with each other, and we should look as much as possible to the other parts to see and understand the whole. This is unfortunately more difficult in the nighttime, as it is impossible to see the creatures who make noises so well as in daylight. However, sharp eyes, a good nose, and a good flashlight (preferably covered with red plastic) will help.

Movements and smells are the two other main means animals use to communicate with each other. With training we can begin to visualize and sense how they are communicating by movement and smell at the same time they are making noises. For example, let us study the overall language of a striped skunk in terms of sight, hearing, and smell. If we hear the angry chittering cry of a skunk in the nighttime, his usual time to be about, we know he is probably communicating in an argumentative way with one of his own kind, because he usually communicates with an enemy of another species, such as a dog or man, by stamping his front feet and raising his tail. This is his way of saying: "Come any closer and I will spray you with something you definitely will not like!" To accentuate this warning, he will often add smell as another communicating signal by letting loose small whiffs of his powerful odor into the air. To a dog who has previously had an unfortunate experience with this smell, this signal is quite enough; he avoids the skunk completely! So, hearing the skunk thumping the ground with his forefeet in the distance creates a mental image of the same animal communicating also by sight and smell warnings.

A skunk's chittering may have different meanings. If the sound is angry, he is probably warning another skunk away from his territory. The chittering could also be a female telling a male to leave her alone. If the noise rises into a regular war cry, it is certain that two skunks are having a fight, and, when the sound stops, one has run away. A softer chittering may mean a mother is calling her children. She may not only turn in different directions to call them, but also let loose a bit of her scent to help guide them to her location. A detectible skunk scent is good evidence of this.

When an animal or a bird is seen in the moonlight or in the red flashlight beam at the same time it is making a noise, observing its actions or detecting its smell will help in understanding what it is saying. Remember to rub a wet finger over your nose to help find a smell on the night air. Two animals growling and walking towards or around each other with a stiff-legged gait are probably two males of the same species each trying to outbluff the other. Often the one who can make himself look big and ferocious enough and also fill the air with a strong smell that indicates his strength can frighten the other into running without starting a fight.

A bird crying or squacking as if in great pain and then moving over the ground dragging a wing is probably trying to lure you or some other dangerous creature away from its young or nest. If an owl hops and twists about on the ground as if it has gone mad, while at the same time uttering blood-curdling shrieks, it is probably a spotted owl trying to lure you away from its nest in its own peculiar manner.

It is said of mammals, birds, and lizards that they fight and communicate more over territory than over mates or food. Test whether this is true by observation. The whippoorwill calling from the roadside hedge in the dusk is probably telling other birds of the same species: "This is my place along here! Keep out!" He shows this further by fluffing up his feathers, even as a dog claiming territory lifts the hair on his back and walks stiff-legged toward his opponent. Thus the whippoorwill flies out at any intruder to chase him away. It is interesting to watch how a bird or animal marks off his territory. Careful watching will make it possible to draw a map after a while showing several territories. Each territory of a separate whippoorwill couple is marked off by the places where the male bird has made his calls all around the edge of the territory.

Foxes, coyotes, wildcats, weasels, and similar carnivorous animals mark their territories with howls, caterwauling, barking and so forth, but much more strongly and permanently mark a territory by leaving scent posts around its edges. The male fox or coyote, like the dog, scratches the ground vigorously where he has just left his droppings or his urine, the deepness of the diggings expressing his size and strength. The male wildcat, lynx, mountain lion, or bear does the same thing by scratching on a tree with his front claws as high as he can reach, making a loud scratching noise while doing so, and then leaving his scent by rubbing his back against the tree.

Most territories are guarded carefully during the mating season and until the babies are reared to near adulthood. Then suddenly, it seems, the animal or bird forgets all about its territory and wanders about where it wills. This is not always true, as a gopher, for example, guards its territory throughout the year. Sometimes also a different type of territory is chosen, after the children have been reared, because certain animals or birds form packs or flocks which have their own special territories. The howling of a wolf pack, or more rarely a coyote pack, will serve as a warning to other wolves or coyotes to stay away from pack territory; and the pack may leave mass smell signs at certain places.

Animals warn each other of danger in four different ways, though not all animals use all four ways. Kangaroo rats, for example, warn of enemies near their desert homes by either striking the ground vigorously with their hind feet to make a loud thumping noise, or by emitting a shrill warning squeak. Similar is the loud smack of a beaver's or muskrat's tail in warning, followed by all nearby beavers or muskrats diving under water. A female fox gives a yipping cry of fear when her cubs are attacked, calling her mate to come help defend them. Then she courageously plunges and snaps at the enemy, even a wildcat or a large dog. But more potent is the long-carrying smell she lets loose, a scream for help in scent form that often brings the dog fox running. If she is killed before help arrives, the pungent smell warns her mate that something terrible has happened and that he must proceed most carefully if he is to save any remaining members of his family. A fourth way of warning is the signal made by the white tails of deer or cottontail rabbits flashing even in darkness to give a warning of danger near, and used not only to warn friends but lead the enemy away from hidden babies.

Watching for these and other means of animal communication will improve one's understanding of the sounds, sights, and smells of the dark, and will greatly add to his pleasure in exploring the night.

Frogs and toads of most kinds depend on their voices to bring together males and females for breeding and the starting of new life when water conditions are best for that life. Being amphibians, they are not protected from drying out as birds, mammals, and reptiles are, but must protect themselves by traveling mainly at night when there is some moisture in the air, living near or in water, hiding in deep holes or caves, or going into a sleep like death underneath slabs of drying mud in desert regions. When the rains start up again and are combined with the proper temperature conditions, frogs and toads instinctively know that it is time to mate and produce eggs before the optimum conditions of moisture disappear again. In the deserts such good moisture conditions are so rare and so short that frogs and toads may have to mate, lay eggs, and have the larvae or young tadpoles grow to adulthood all in a period as short as two weeks! No wonder the time of a rain in the desert is signaled by the loud, frantic cries of the males at any temporary pool that is forming. In effect, they are saying at least subconsciously to the females: "Hurry up! We have only a few hours to prepare the eggs!"

Whenever and wherever warm rains come after a period of dryness, conditions are optimum for going out into the darkness and watching many a scene of frantic activity among amphibians. Male frogs and toads come hopping through the woods, prairies, meadows, brush, grasslands, or deserts, splashing happily in the small puddles but looking for the pools that will be big enough for the laying of eggs. Soon the choruses in all the extraordinary variety described in the following pages begin to sound through the darkness, everything from shrill peeping to deep-throated roars. Soon the females begin to follow these siren songs through the gloom to where the males in the pools are inflating their throats in various ways to send out their far-carrying cries. The male frog or toad clasps with his front legs whatever female lands in the water nearby; eventually, the milky milt of the sperm is poured out over the eggs in the water, and a jellylike substance made by the female covers and protects them until they are ready to hatch.

One of the most extraordinary spectacles to watch in the night is the male frog's or toad's throat expanding into a regular little balloon as the thin skin is stretched like rubber until you would surely think the large bubble it produces would break! In some species the swelling pouch appears in the center of the throat, in others low down or high; or perhaps two bubblelike extensions appear at once, one on each side of the head. Whatever way it happens, plenty of noise is produced and the wild racket lasts through most of the night, dying finally to nothing when the last egg is laid and the new generation is safely launched on the water trail of frog and toad life.

Key to Using the Guide to
Bird, Insect, and Animal Sounds

The Guide to Bird, Insect, and Animal Sounds, which follows, divides the sounds into various categories to help the reader pinpoint whatever sound he may be hearing at the moment. The Guide begins with bird songs and calls divided into the following categories: Loud or Startling, Medium to Soft Songs, and Semimusical Songs. It continues with musical bird and amphibian calls, classified as follows: Whistling, Nonwhistling, Agitated Semimusical Calls, Soft Semimusical Calls, and Other Semimusical Calls. Next, the Guide describes those sounds of birds or insects which are definitely nonmusical, divided into the following classifications: Wing Sounds, Mechancial- or Metallic-sounding Noises, and Noises That Are Not Mechancial- or Metallic-sounding. Then the sounds of mammals are described, arranged in these subdivisions: Grumbles, Growls, Roars, Snarls, and Squalls; Bellows, Grunts, and Coughs; Loud Squeals and Screams; Yelping and Barking; Yowls and Howls; Chattering, Scolding Calls; Squeaks and Other Shrill Calls; Noises Animals Make with Their Bodies, and Water Sounds. The Guide concludes with the sounds made by creatures in chorus under the following headings: Musical Choruses of Amphibians and Birds, Nonmusical Amphibian Choruses, and Musical Insect Choruses.

GUIDE TO BIRD, INSECT, AND ANIMAL SOUNDS

BIRD CALLS, NOTES, SONGS, AND HOOTS*

Mockingbird

	Description of Sound	*Source of Sound*
Loud or Startling	Extremely varied song, often copying other birds, but also with liquid as well as rasping harsh notes of its own. Heard mainly during summer nights, especially in moonlight.	Mockingbird
	Squealing whistle in forested areas.	Male long-eared owl calling to female
	Resonant, deep, hollow "hoo-hoo, hoo-hoo, whooo!", more rapid in winter (mating time).	Great horned owl
	Booming "whoo-oo-oo-oo."	Great gray owl
	Deep hoarse croaking.	Great blue heron
	A little less deep croak.	Little blue heron
	Rough guttural note, like "quaw-quaw."	Green heron, usually at mating time
	Explosive alarm note of "Quoh!"	Least bittern
	Loud but not deep "chok!" or "quack."	Black- or yellow-crowned night heron
	"Hoos" in 2 groups of 4-5 syllables, as "hoo-hoo-to-hoo."	Barred owl

Long-eared Owl

* Include sounds made by creatures whose calls resemble those of birds.

Great Horned Owl

Least Bittern

Black-crowned Night Heron

Great Gray Owl   **Great Blue Heron**   **Little Blue Heron**   **Green Heron**

Weird noisy hoots and screams mixed, forceful and swinging rhythmically.

Barred owl

Resonant, crazy laughter, part screaming.

Common loon

Strident honking.

Geese, generally on water, when disturbed at night

Barred Owl

Harsh, cackling cry, "cak-cak-cak-cacaha, caha!", somewhat like chicken.

Clapper rail

Squacking cries.

Ducks

Squacking "yak-yak-yak" cries.

Young black-crowned night herons

Harsh, discordant scream of alarm.

Barn owl

Piercing shriek of warning in the forest.

Long-eared owl, when its young are being approached

Loon

Shrill scream.

Burrowing owl, when attacked or wounded

Harsh whistles, "queedle" or "queeee."

Golden plover

Piercing scream or yoddle.

Greater yellowlegs

Piercing "wheeeeep!" whistle, often repeated rapidly, heard on ocean and bay shores.

Black oyster catcher

Blaring screech.

Barking frog, when attacked

Harsh "kur-lee."

Long-billed curlew

"Peent" or "beedz!" from up in sky.

Common nighthawk

Piercing "speek-speek" call, from sky.

Common nighthawk

Harsh "kret-kret" call.

Marbled godwit

Clapper Rail

Marbled Godwit

Long-billed Curlew

Nighthawks

Barn Owl

Barking Frog

Greater Yellowlegs

Black Oyster Catcher

Golden Plover

Burrowing Owl

Killdeer

Wilson's Snipe

Willet

Swainson's Thrush

Field Sparrow

Purple Martin

	"Kill-dee," "kill-dee-dee," or "dee-dee" call, heard either on ground or in sky at night.	Killdeer
	Harsh "escape, escape" cry.	Wilson's snipe
	"Cay-ti" call, with second syllable lower.	Willet
Medium to Soft Songs	Very sweet song, spiralling upward, heard usually in evening and on moonlit night.	Swainson's thrush
	Clear tenor chant of "he-ew, he-ew, he-ew, hewhew-hew, he-eeeu," accelerating in spiral sound; heard just before dawn.	Field sparrow
	Light, clear, carolling, rising and falling song of "fill-em, air-em, giv-em, musik"; heard during spring and summer in late evening, moonlight, or early dawn.	Robin
	Gurgling song, changing to rich and throaty "tehew-wew" or "pew-pew" notes; heard on moonlit nights or in late evening.	Purple martin
	Bold but disconnected phrase of clear notes, welling up into 3 very beautiful flutelike notes, then a trill; heard in late evening or on moonlit night.	Wood thrush
	Long clear opening phrase, flutelike in tone, followed by 3-4 other beautiful flutelike notes; heard in late evening or on moonlit nights.	Hermit thrush
	Ghostly, disembodied spiral note of continuous tone, rising from darkness, then sliding down the scale with a double chord; heard in late evening or on moonlit nights.	Veery
	Melodious, floating song with water-gurgling notes, followed by "wheep-ee-ee-you!"	Upland sandpiper
	"Peet-peet-, pee-ter-wee-too, peetey, wee-to, we-too-wee-too," coming down from sky at dusk.	Long-billed dowitcher, at mating time
	Bits of music like water dripping, growing louder and then falling away, ending in very long trill; sometimes heard on moonlit night.	Wood wren
Semimusical Songs	Various odd harsh and whistled notes in repeated series, with long pauses between series, like "wek-wek-wek-wek-wek-wek; kook-kook," etc.	Long-tailed chat
	Reedy series of notes, interspersed with gurgling notes; heard in marshes on moonlit nights.	Marsh wren
	Metallic gurgling "onk-la-reeee," ending in quaver; heard in marshes.	Redwing blackbird
	"Sweet-sweet-sweet," followed by both buzzing and musical notes; heard in moonlight.	Song sparrow

Screech Owl

Black-bellied Plover

Bobwhite

Great Gray Owl

Wood Thrush

Hermit Thrush

Great Horned Owl

Snowy Plover

Pygmy Owl

Veery

Wheezy notes combined with ventrilo-quial whistle for 1-2 seconds, lowering in pitch; heard in southwestern desert country after rains.

Sonoran green toad

MUSICAL BIRD AND AMPHIBIAN CALLS*

Long-billed Dowitcher

	Description of Sound	*Source of Sound*
Whistling	"Whurdle-whurdle-whurdle" in evenly spaced series; also clear staccato whistle repeated at half-minute intervals.	Saw-whet owl
	Tremulous, often weirdly quavering whistle, with wailing, whinnying notes, rising in pitch, then sliding down into quiver.	Screech owl
	Plaintive, slurred "wheeee-ee-eee."	Black-bellied plover
	Tremulous, vibrating whistle like that of screech owl, but deeper.	Great gray owl
	Musical whistle of "ooo-hee."	Great gray owl
	Musical whistle of "ooo-hi."	Great horned owl
	Clear "bob-white" call in spring.	Bobwhite
	Peculiar soft pleasing whistle from deep woods, often repeating "too-too-too-tuk-tuk," etc.	Pygmy owl
	Single hollow note early in morning.	Pygmy owl
	Low-pitched "pik-pik-kroor-ree" or "pee-wee-uh-oo-weeah!"	Snowy plover

Song Sparrow

Saw-whet Owl

*Many of the calls listed under Musical Choruses of Amphibians and Birds are sometimes given alone without forming choruses. Study these descriptions too, as they are not duplicated in this section. Also, the calls contained in this section include the calls of other animals whose voice sounds resemble those of amphibians and birds.

Mountain Plover

Semipalmated Plover

Long-billed Curlew

Striped Skunk

Low, variable whistle.	Mountain plover
Pearl-toned, clear "tchi-wee" or "too-wee," rising in pitch.	Semipalmated plover
Rapid "klee-le-le-lee" call.	Long-billed curlew
Cooing whistle followed by purr.	Male striped skunk courting female

Nonwhistling

Long trilling note.	Pygmy owl at courtship time
Clear, musical "pill-will-willet."	Willet
Soft, hollow-voiced "too-too-too-too-too; toot; toot; toot," with pauses between last three notes.	Pygmy owl
Cooing like that of mourning dove, but sharper and more distinct.	Pygmy owl
Mellow chirp, sounding a protesting note.	Western toad, when handled
Clear chirping notes heard from thick trees at night.	Flying squirrel
Deeper chirping call, heard mainly from ground levels and near large piles of brush or cacti, etc.	Wood rat
Clear trilling heard mainly from ground areas.	White-footed mouse
Chirp heard in low areas near waterways.	Otter
Soft trill from low on ground, heard in high Sierras.	Yosemite toad
Clear trill, lasting 7-10 seconds, rising in pitch and ending suddenly; heard on damp nights.	Southwest toad
Even-pitched, clear trill, lasting 6-10 seconds; heard along streams and in arroyos and washes, especially on damp nights.	Red-spotted toad
Sweet, plaintive "er-eeeee," with rising inflection, coming from marshes.	Sora rail
Delightful, happy, chirping call at night, from deep woods.	Richardson's owl

Willet

Wood Rat

River Otter

Sora Rail

Western Toad

Flying Squirrel

White-footed Mouse

Yosemite Toad

Low-pitched "kee-hou-kee-hou." — Bobwhite coveys in brush

Liquid peeping heard in woods. — Young of saw-whet owl calling for food

Trill or chirp similar to cricket's, but more froglike, coming from rocky areas. — Cliff frog or Rio Grande frog of lower river

Soft, mournful "coo-coo-coo; coo; coo," with last note softer. — Mourning dove

Mellow, sonorous, far-reaching "cooo-oo-oo-oo." — Love song of burrowing owl

Agitated Semimusical Calls

Rapid "chi-le-le-le, chi-le-le-le." — Young barn owl calling for food

Rapidly repeated series of "toot-toot-toot-too," etc. — Short-eared owl calling for mate

Rapid, excited "whup-whup-whip-whirr." — Young whippoorwill in distress

Fighting, angry "cou-wee, wou-wee," varying to "huah-hua" and "whe-ah-hua." — Male ruffed grouse

Sharp-sounding, rapid "ho-ho-ho-ho." — Screech owl

Rapid peeping or squeaking. — Young screech owls

Emphatic whistled "wit" or "weep." — Wilson's plover

Sharp "week" or "keep," excitedly repeated. — Avocet

Rapidly repeated "whee-doodle, whee-doodle." — Greater yellowlegs

Repeated "wee-wee-wee-wee-wee," not so sharp as call of greater yellowlegs, heard from sky. — Wandering tattler

Rapid, metallic, treble-tone "to-tu-tu-tu," often from sky. — Dowitcher

Strident "ka-reee" when heard close at hand, and several short, sharp cries, coming from marshy areas. — Sora rail

Noisy, repeated "coo-coo-coo-coo-coo." — Coot

Bobwhite

Burrowing Owl

Short-eared Owl

Ruffed Grouse

Sora Rail

Coot

American Avocet

Greater Yellowlegs

Wandering Tattler

Screech Owl

Long-eared Owl

Chuck-Will's-Widow

Saw-whet Owl

Bobwhite

Screech Owl

Poorwill

Nighthawks

Woodcock

Colorado River Toad

Short-eared Owl

Soft Semimusical Calls	Hooting of "quoo-quoo-quoo" or "quoo-oo-oo," coming from deep woods.	Long-eared owl
	"Whoof-whoof-whoof," heard in deep woods.	Female long-eared owl
	"Tu-wink, to-wink."	Long-eared owl
	Peculiar soft quavering cries, heard in open fields and marshes.	Short-horned owl
	"Poo-woo," "poor-will-low-low," or "puih-we-he," heard just before dawn or in late evening, usually in brush.	Poorwill
	Guttural purring notes, heard in open areas.	Female nighthawk calling to young
	Childlike "keeeer."	Young screech owl calling to parents
	Single melancholy note, repeated every 1-2 minutes, in addition to a high-pitched but soft "ting; ting; ting; ting" like the tolling of a bell; heard in deep woods.	Richardson's owl during courting
	Low, slurred "peent," sounding like bird trying to get mud out of bill; heard on warm, moonlit nights.	Male woodcock
	Bass toot, like distant ferryboat whistle; heard in or nearby streams.	Colorado River toad
Other Semimusical Calls	Squealing as if in pain.	Short-eared owl luring intruder from nest
	"Chuck-will's-widow," repeated over and over, sounding from afar like "puck-whee-ee," in addition to "poor-will-unk" at dawn or dusk; heard in riverbottom country in South.	Chuck-will's-widow
	"Wheatchie," repeated in groups of 4-5.	Bobwhite
	"Ka-lo-i, he-e" or "hurl-le-e, he-e," uttered in series of 4-5 at about 1-second intervals.	Bobwhite
	Vibrant, penetrating "whoop" or "kwook," about 3 every 2 seconds; usually heard on full-moon nights in woods.	Male saw-whet owl
	Tremulous, lugubrious wailing from woods or brush.	Screech owl

Burrowing Owl

Pygmy Owl

Semipalmated Plover

Sora Rail

Solitary Sandpiper

Spotted Sandpiper

Wilson's Snipe

"Whit-whit, who-who-whit."	Burrowing owl
Long whistle followed by cuckoolike "cuck-cuck-cuk."	Pygmy owl
High-pitched, whinnying note, often mixed with some whistling.	Pygmy owl
Series of queer throaty "peents."	Male woodcock courting female
Sharply marked "peet-weep" with first note higher; heard in late evening.	Solitary sandpiper
Plaintive "chee-wee" or "tooo-eee," with second note higher.	Semipalmated plover
Strangely beautiful call like tinkling of ice in glass of water; heard from sky.	Red-backed sandpiper
Sharply uttered "peep-weep," dropping in tone.	Spotted sandpiper
Long whinnying call, coming from marshes.	Sora rail
Whinnying call, given while bird is making peculiar wobbling flight.	Wilson's snipe
Part dovelike, part froglike "Uh-uh-uh-uh-oo-oo-ooo-oooh."	Male least bittern cooing to female
Chirping and squeaking low to ground in desert.	Geckos, various mice, and kangaroo rats

Least Bittern

Kangaroo Rat

NONMUSICAL SOUNDS OF BIRDS OR INSECTS, ETC.†

	Description of Sound	*Source of Sound*
Wing Sounds	Loud "zum-zum-zum."	Male turnstone courting female in sky
	Sound like "zoom-zoom" or "who-who-who-who."	Wilson's snipe giving spring mating call by directing wind through feathers in special way when flying up in sky

Black Turnstone

† Sounds include those of some mammals, reptiles, and amphibians that sound like birds or insects.

Woodcock

Short-eared Owl

Bobwhite

Killdeer

Glossy Ibis

Nighthawks

Whippoorwill

	Chattering.	Woodcock whirring wings when flushed; sound made with two stiff narrow feathers of outer primaries
	Buzzing.	Pygmy owl flying low to ground
	Clapping.	Short-eared owl
	Whirring.	Male bobwhite quail, when flushed or issuing challenge to combat
	Loud "whoosh."	Killdeer or willet, when intruder gets too near nest
	Deep humming.	Flock of glossy ibis diving out of sky
	Deep, resonant, boomlike "swrrrr-coonk."	Male nighthawk at bottom of high dive from sky
	Deep, almost roaring boom.	Ruffed grouse bringing wings toward body at high speed and stopping them suddenly
	Soft whirring.	Hawkmoths hovering at night flowers
Mechanical- or Metallic- sounding Noises	Loud snapping of bill.	Large owl, when angry or courting
	Medium-noisy snapping of bill.	Medium-sized owl, when angry or courting
	Soft snapping of bill.	Small owl, when angry or courting
	Rasping and sucking	Young barn owls
	Sound like corn-popping.	Whippoorwill while courting
	Rapid "click-clack; click-clack."	Nighthawk young in flight
	"Cac-cac, cac, cac, ca-ca, caha, cah."	Clapper rail, when disturbed
	Clicking or ticking "kik-kik-kik-ki-queah," last note with rolling sound.	Yellow rail
	Notes like saw-filing, or metallic "tong-tong-tong."	Saw-whet owl or coot

Ruffed Grouse

Clapper Rail

Saw-whet Owl

Stridulating metallic-sounding warning.	Velvet ant and other ants
Rattling noises.	Mandibles of burrowing owl
Rattle or buzz like an explosion of steam.	Rattlesnake rattling its tail
A truer rattle.	Tail of gopher, bull, copperhead, or king snake in dry leaves
Rattling noise heard in big pile of brush or in abandoned cabin among old papers.	Wood rat giving a warning by vibrating his tail
Rattling of sticks and cones.	Porcupine
Low-pitched rattling sound.	Pacific giant salamander
Rattling of leaves.	Claws of armadillo
Combined buzzing and trilling call.	White-footed mouse
Low buzzing call.	Whippoorwill
Underground buzzing.	Rapidly digging gopher
Weak buzz like locust.	Twin-spotted rattlesnake
Buzzing like swarm of bees.	Chorus of narrow-mouthed toad
Loud buzzing, clattering noise.	June beetles
Popping, purring noise.	Male whippoorwill in brush
Popping sound.	Coral snake or hognosed snake suddenly everting the vent in its tail
Similar popping sound combined with bubbling noise.	Western hog-nosed snake, when attacked
Sandpaper-like sound.	Chuckwalla or other lizard sliding into rock crevice

Mutillid

Burrowing Owl

Western Diamondback Rattlesnake

Wood Rat

Chuckwalla

Arizona Coral Snake

American Porcupine

Narrow-mouthed Toad

June Beetle

Whippoorwill

White-footed Mouse

Gopher Snake

Pacific Giant Salamander

Pocket Gopher

Poorwill

Bobcat

Long-eared Owl

Badger

Least Bittern

Opossum

Chuck-Will's-Widow

American Bittern

Beaver

River Otter

Blowing noise.	Poorwill, when threatened or wounded; or hog-nosed snake, known as "puff adder," puffing up to scare attacker
Hissing noises.	Most snakes (volume usually depends on size); also, red fox pups when alarmed
Hissing "haah!"	Least bittern
Short whistling, hissing, whoostling sound.	Porcupine, when attacked
Loud hiss.	Opossum; alligator lizard; barn owl; barred owl
Angry hissing note.	Chuck-will's-widow; common nighthawk
Very loud hiss.	Large alligator
Wheezing hiss.	Gopher attacking another gopher
Hissing snarl.	American bittern
Hissing combined with swishing in water.	Beaver
Light hissing.	Weasel
Hissing from streams or lakes.	Otter or mink
Spitting and hissing.	Bobcats, Canadian lynxes, and mountain lions
Spitting-hissing noise.	Long-eared owl
Asthmatic wheezing hiss.	Badger
Rasping, sibilant "tzzzzzzzz" like jets of steam.	Young saw-whet owl
Soft murmuring, heard mainly in open woods and meadows.	Red fox calling pups out of den
Kittenlike mewing.	Kittens of bobcats, Canadian lynxes, and mountain lions; also red fox pups
Whimpering, crooning, mumbling, and murmuring cries, often together; also murmuring "coo-coo-coo," softer than dove.	Porcupine

Noises That Are Not Mechanical- or Metallic-sounding

Red Fox

American Porcupine

Flying Squirrel

White-tailed Deer

Striped Skunk

River Otter

Sora Rail

Coyote

Mumbling from water area.	Beaver
Soft chirring.	Flying squirrel
Soft chirring, twittering from ground.	Female skunk calling young
Soft chuckling.	Male otter in courting
Snuffling in water.	Otter
Purring.	Bobcats, Canadian lynxes, and mountain lions; also young weasels and mink
Low, whining murmur.	Sora rail, when enemy approaches nest
Soft bleating.	Female deer calling to young
Whimpering and whining.	Hungry pups of red fox, coyote, wolf, weasel, mink, otter, wild dog, gray fox, and desert fox
Moaning.	Red fox female protecting young
Louder whining.	Female coyote calling young to eat
Pitiful or high-pitched whining.	Male porcupine calling to mate
Deep, loud moan.	Black bear, when hurt
Low whining, moaning, and whimpering combined.	Black bear cubs, when left alone
Whimpering.	Young cottontail or brush rabbits, young beavers, and baby barn owls
Whimpering and wailing.	Hurt beaver
Muffled whimpering.	Young ring-tailed cats
High whimpering cry.	Mountain beaver
Angry guttural notes.	Great horned owl
Nasal, grating caterwauling.	Ruffed grouse sounding battle call

Black Bear

Mountain Beaver

Great Horned Owl

Ruffed Grouse

Saw-whet Owl

Wilson's Phalarope

Wilson's Snipe

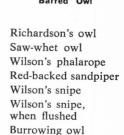

Barred Owl

Burrowing Owl

Whippoorwill

Chuck-Will's-Widow

Canada Goose

Yellow-crowned Night Heron

Peculiar grating call.	Richardson's owl
Rasping, querulous "saaay."	Saw-whet owl
Nasal "wurrk."	Wilson's phalarope
Nasal, rasping "kzeeep."	Red-backed sandpiper
Low, guttural "skiap."	Wilson's snipe
Nasal rasping note.	Wilson's snipe, when flushed
Raucous "twit-twit-twit-twit."	Burrowing owl
Nasal "ooh-ick-oooh-ick."	White-faced glossy ibis
Clucking, clicking, squeaky "crick-a-crick," heard in woods.	Barred owl
Clucking notes.	Chuck-will's-widow or poorwill, when flying low
"Gluck."	Female whippoorwill during courting
"Cack-cack-cack-cack."	Burrowing owl, when alarmed
Honking from sky or around lakes, ponds, etc. at night.	Canadian and other geese
Quacking from sky or around water at night.	Ducks in migration or when disturbed
Quacking noises from bees' nest.	Immature queen bee making short calls, increasing in length, announcing imminent attack on main queen
Quacking "yak-yak-yak," in trees near water.	Young of black-crowned night heron
Gabbling cries from lakes, marshes, etc.	Geese and ducks
Clucking heard in coniferous woods and meadows.	Female marten at mating time
Quacking loud notes.	Yellow-crowned and black-crowned night herons
Nasal clucking around ponds, marshes, etc.	Coot

Marten

Coot

Queen Bee

Purple Gallinule

Sora Rail

Bobwhite

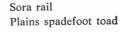

**Black-crowned
Night Heron**

Long-eared Owl

Burrowing Owl

Black-necked Stilt

Red-legged Frog

Barn Owl

Less nasal clucking around ponds, marshes, etc.	Gallinule
Clucking "cut-cut-cut-cutta."	Sora rail
Ducklike "quack," lasting about 1/6 second, heard in open ground.	Plains spadefoot toad
Peculiar buzzing grunt.	Whippoorwill
"Wak," followed by squeal.	Black-crowned night heron giving warning
Shrill prolonged squeak from trees in woods.	Long-eared owl
Squealing whine.	Ruffed grouse, when defending young
Crowing or caterwauling "hoo-poo-weih-hurrah-ha-he-weil, wah, weh-ha-hah" on moonlit nights.	Bobwhite quail at mating time
Same call heard in early dawn.	Roadrunner
Wild "kah-kaw-kaw."	Male and female burrowing owls, sitting on burrow mound at mating time
Chattering call.	Burrowing owl, when flying about
Cries with yapping notes.	Black-necked stilt, when flying
Tooting from bees' nest.	Queen bee protesting when workers try to stop her from killing young queen
Long, anxious, bleating "yee-oow," dropping in pitch.	Couch's spadefoot toad
Chuckling call, heard in groups of 3 to 4 near or in placid water.	Red-legged frog
Hoarse "wah-wah-wah-wah" rapidly repeated, each about ¼ second long.	Great Plains spadefoot toad
Hoarse snoring call, with ratchetlike sound similar to stroking comb teeth with fingernail.	Western spadefoot toad
Nasal, snoring, wheezy, explosive call of "waaa-a-a-ah," halfway between a baby crying and a calf bawling.	Woodhouse's toad
"Ik-ik-ik-ik."	Barn owl calling in flight to signal that food is coming

Long-eared Owl

Poorwill

Nighthawks

Ruffed Grouse

Squeaking cry of "crick-o-crick" from trees in woods.	Long-eared owl
Shrill, prolonged squeak from trees in woods.	Long-eared owl
"Poor-will-uck" at early dawn or late dusk.	Poorwill
"Dik-dik-dik-dik-dik."	Male common nighthawk chasing another in sky
Squealing whine.	Ruffed grouse, when defending young
"Quit-quit-quit-quit-quit."	Ruffed grouse, gathering its young; also various quail
Squeaking from off ground.	Bats; also saw-whet owl
Chattering, friendly call.	Young saw-whet owls in nest, when eating
Sharp "whit" or "kit."	Red or northern phalarope
Clear, 3-syllabled "hew-hew-hew" or "deer-deer-deer," heard on shores.	Greater yellowlegs
"Ku-ku-ku" or "whea," heard on shores.	Lesser yellowlegs
Thin, mouselike or pipitlike "jeet," heard along shores.	White-rumped sandpiper
Short, distinctive "kip" or "kit," heard along shores.	Sanderling
Sharp, drawn-out, thin "creeee-eeeet," heard along shores.	Least sandpiper
Low "tut-tut-tut" or "uk-uk-uk."	Least bittern
Call of "tell-u-what!"	Little blue heron
Harsh alarm cry of "skeow!"; peevish cry of "fly-up-the-creek."	Green heron
"Creek-creek-creek."	King rail on ground

Saw-whet Owl

King Rail

Little Blue Heron

Green Heron

Red Phalarope

Least Bittern

Least Sandpiper

Northern Phalarope

Greater Yellowlegs

Lesser Yellowlegs

Sanderling

"Cark-cark-cark." | King rail in flight
"Churck" cry. | King rail, when disturbed

Hoarse croak. | Great blue heron; little blue heron

Croak of disgust. | American bittern
Growling, guttural croak. | Yellow-legged frog
Croaking at night from trees. | Flocks of crows and ravens, when alarmed

Great Blue Heron

SOUNDS OF MAMMALS*

Description of Sound | *Source of Sound*

Grumbles, Growls, Roars, Snarls, and Squalls

Snarling, growling, and squalling. | Attacked or attacking red fox, gray fox, kit fox, mountain lion, Canadian lynx, bobcat, bear

Growling and snarling. | Fighting mountain lion, Canadian lynx, bobcat, coyote, wolf, bear, raccoon, weasel, marten, mink, otter

Growling and hissing. | Fighting badger, weasel, marten, mink, skunk, otter, etc.

Very loud squalling and bawling. | Black bear, when frightened or in pain

Grumbling, teeth-clicking noise. | Porcupine, when attacked

Growling and squealing. | Fighting woodchucks and marmots

High-pitched growl. | Shrews
Mumbling growl. | Male jackrabbit in fight
Powerful bugle call, starting as roar and ending with explosive scream. | Bull elk

Deep note, often turning into shrill squeal. | Young bull elk

American Bittern

Common Crow

Red Fox

* Sounds of some birds, reptiles, amphibians, insects, etc. which resemble those of mammals are included.

Mountain Lion

American Porcupine

Badger

Woodchuck

White-tailed Jackrabbit

Canadian Elk

Raccoon

Wild Boar

Long-tailed Weasel

Red Fox

California Sea Lion

Black Bear

River Otter

Moose

Bullfrog

Bellows, Grunts, and Coughs	Loud roaring.	Fighting black or grizzly bear
	Churring growl.	Raccoon
	Snoring growl.	Badger
	Roaring noise on beaches.	Sea lion bulls
	Comparatively soft grunting.	Porcupine or armadillo
	Coughing grunt.	Black bear, when about to attack
	Grunts.	Female black bear ordering cubs to climb tree when danger is near
	Grunting near water.	Otters at play
	Grunting in open or brushy areas.	Male jackrabbits fighting each other
	Bellows and grunts.	Bull elk fighting another bull
	Angry bellowing and grunting combined.	Moose
	Low, challenging grunts.	Male moose
	Deep, savage grunts.	Badger
	Bass bellowing sounding like "jug-o-rum" from pool.	Bullfrog
	Throaty, bellowing roar.	Bull alligator
	Moaning grunts like "umph-umph-umph."	Young alligators with mouths closed
	Snarling, explosive grunt.	River frog
	Guttural grunts.	Wild pigs, wild boar, or pig frog
Loud Squeals and Screams	Squeals.	Young male elk
	Shrill screams.	Most hares and rabbits, when attacked or wounded
	Snarling screams.	Injured weasel, marten, fisher, skunk, wolverine, otter, or mink
	Blood-curdling shriek.	Red fox
	Scream like a badly hurt woman.	Female mountain lion at mating time
	Squalling screams.	Lynx or bobcat
	Piercing scream.	Ring-tailed cat

Tree Frog

Canadian Lynx

Barking Frog

Bobcat

Wild Horse

White-tailed Deer

Red Fox

Coyote

Mink

Gray Fox

	High-pitched, squacking scream.	Green frog jumping into water to escape enemy; or large female tree frog, when hurt or seized
	Incredible screams, yowls, and cater-wauling.	Bobcat and lynx, when fighting or mating
	Screeching scream, usually from up in trees.	Tree squirrels
	Whinnying scream and snorts.	Wild horse stallion, when attacking
	Whistling snort.	White-tailed or mule deer, when alarmed
	Savage snorts.	Male deer or wild horse stallions, when attacking other males
Yelping and Barking	Sharp, terrierlike bark; also 2-3 yaps, followed by stirring "yaaaaaaaa!"	Red fox
	"Kalyaks" and loud barks.	Female red fox, alarmed by intruder approaching young
	Light yapping call.	Red foxes communicating about game
	Yaps and yelps together.	Male and female foxes conversing
	Coughing bark.	Fox warning of danger
	High-pitched yaps.	Coyotes, especially young
	Deeper-pitched bark.	Coyote or wolf giving warning
	Barking on beaches.	Sea-lion pups, and sea otter
	Yipping bark, heard in coniferous forests.	Porcupine
	Very shrill barks interspersed with coughlike bark.	Ring-tailed cat
	Snappy bark, usually heard near water.	Mink
	Coarse, croaking yelp.	Gray fox
	Series of explosive barks, given at intervals of 2-3 seconds, usually heard near limestone caves.	Barking frog

Black Bear

Red Squirrel

Wood Rat

River Otter

Yowls and howls

Chattering, Scolding Calls

Squeaks and Other Shrill Calls

About 8-10 raucous barks heard from high in tree.	Barking tree frog
Sharp howl.	Female red fox calling for help when protecting young
High-pitched, mournful, haunting howl, followed by series of sharp yips and yaps, often ventriloquial and sounding like a chorus of a dozen or more, when only a few are howling.	Coyotes
Deep bass howl with bawling note.	Black bear, in fright or pain
Deep bass mournful and very wild-sounding howl.	Timber wolf
Many mixed kinds of lesser howls.	Wild dogs
Deep-toned yowling howl.	Large male bobcat or lynx
Shrill, threatening, chattering noise.	Female shrew rejecting male
Frantic, scolding chatter.	Red, gray, and fox squirrels warning of danger
Scolding cry.	Wood rats
Chattering call in water.	Otter playing around water
Squeaking, often heard from large hole in ground.	Young coyote in distress
Shrill squeaking cries from up in air.	Bats
Constant, chittering squeal.	Myotis bats, when hanging or flying in caves, barns, etc.
Very shrill whistle.	Male mountain lion responding to female
Loud squeaks in forest.	Porcupine
Strange falsetto cries.	Male porcupine at mating time
Faint squeaks.	New-born rabbits, mice, or rats
Shrill, high-pitched, growling squeak.	Attacking shrew
Angry squeak.	Female shrews, mice, and rats rejecting males, or males attacking other males
Crying like hurt child.	Hurt or lonely beaver or bear cub
Short, sharp squeals, like "chuck-chuck."	Flying squirrel expressing alarm

Long-eared Myotis

Mountain Lion

American Porcupine

Flying Squirrel

Clear, musical, chirping note.	Flying squirrel	
An exchange of shrill squeals.	Wood, Norway, black, cotton, and other rats, when fighting	
Shrill chirps and crowing.	Baby weasel	
Squealing cries.	Raccoon young calling for help or food	**Grasshopper Mouse**
Long shrill squeak sounding like a hunting howl.	Grasshopper mouse	
Squeaks and chirps from desert or semitropical areas.	Geckos	
Squeaks.	Arboreal salamander, when caught	
		Banded Gecko

Noises Animals Make with Their Bodies

Thrashing and swishing.	Porcupine defending itself with sharp-quilled tail	

Loud chattering of teeth.	Porcupine or wood rat, when disturbed	**Arboreal Salamander**
Grinding noise.	Marmot or woodchuck grinding teeth when angry	
Rustling noise.	Feet and tail of wood rat in dry leaves or paper in old cabin	
Gnawing noise.	Porcupine chewing on bark or salty wood; or beaver cutting down tree	
Gnashing of teeth.	Coyote, wolf, or fox, when angry	**Yellow-bellied Marmot**
Rustling noises along stream at night.	Raccoon hunting food	
Ripping sound of scratching on trees.	Male badger, bear, bobcat, lynx, or mountain lion clawing trees to sharpen claws or show its own size; or wild boar or peccary scratching trees with its tusks	
		Raccoon
Scuttling noises on beach.	Rock crabs scurrying under rocks	
Scratching sound in leaves.	Box turtle, tortoises, armadillo, wood rat, and opossum	
Rattling of tin cans in garbage can at night.	Bears, opossums, raccoons, coyotes, dogs, and armadillos	**Badger**
Scratching in trees.	Opossum, raccoon, tree squirrels, marten, fisher, black bear (especially cubs or juveniles), bobcat, wild domestic cat	
Digging, rooting noise.	Wild pigs digging up ground for tubers, roots, nuts, etc.	**Opossum**

Ground Beetle

Cottontail Rabbit

Wood Rat

Striped Skunk

Mule Deer

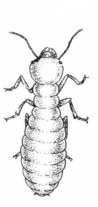

American Bittern

Moose

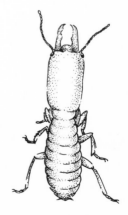

**Western Termite
(Worker)**

**Western Termite
(Soldier)**

Light scratching rustling noises in dry leaves.	Large ground beetles
Soft scratching, digging noises.	sexton beetles burying small dead animal or bird
Thumping on or in ground.	Cottontails, hares, and jackrabbits thumping ground with hind feet to warn of danger; kangaroo rat thumping ground in burrow before emerging
Lighter thumping on ground.	Wood rat thumping ground with hind feet to warn of danger
Thumping on ground, combined with light skunk smell.	Skunk thumping ground with forefeet **to** warn trespasser of impending attack
Loud, thumping, crashing noise.	Mule deer striking ground with all four feet
Thumping noise like large stake being driven into marsh.	American bittern clapping its wings
Crashing sounds in brush.	Deer, moose, elk, and bear
Faint clicking from inside wood, audible by putting ear to it.	Termites and carpenter ants

Click Beetle

Western Diamondback Rattlesnake

Grizzly Bear

Pistol Shrimp

Arizona Coral Snake

Beaver

Muskrat

Nutria

American Avocet

Horned Grebe

River Otter

Clicking behind wall of home or building.

Clicking in grass or plants.

Harsh, explosive clatter of 5-50 seconds.

Soft clicking as of castanets.

Thundering sound.

Explosive snapping heard from tide pools of ocean shores.

Popping heard on beach after a long, hot day.

Cracking and popping.

Popping.

Water Sounds

Distinct, explosive spatting.

Combined spatting and plunking.

Plunking, slapping noise.

Solid plunk like large rock hitting water and sinking.

Swishing plunk.

Lighter plunk.

Swishing.

Bat climbing along wall

Click beetles

Great Plains toad

Rattlesnake clicking rattles when slightly disturbed

Grizzly bear rolling boulders

Pistol shrimp snapping pincers

Seaweed balls

Crabs snapping large claws

Hog-nosed snake or coral snake opening its vent suddenly

Beaver slapping water with broad, flat tail to warn of enemies near

Muskrat slapping water with partly flattened tail to warn of enemies near

Nutria slapping water with round tail

Turtle falling into pool

Beaver, otter, muskrat, mink, nutria, or seal diving; or large fish leaping

Small to medium fish leaping; crab dropping into tide pool; frog dropping into pool

Avocet sweeping water of shallow pool with bill to stir up prey; grebe, loon, duck, otter, mink, muskrat, beaver, nutria, or seal diving from top of water

Mallards

Swishing and splashing.

Careless swimming of otter, mink, muskrat, nutria, seal, beaver, or ducks (especially those that duck heads under water for food and paddle with feet); otter sliding down mudbank into water

Mink

Soft hissing and swishing.

Otter, mink, muskrat, beaver, nutria, seal, duck, or goose swimming almost noiselessly to avoid detection by enemy

Splashing.

Feet of coot, loon, grebe, or diving duck hitting water as it takes off from lake or pond

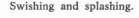

MUSICAL CHORUSES OF AMPHIBIANS AND BIRDS*

Coot

	Description of Sound	*Source of Sound*
Whistles, Whistling Trills, or Toots	High pipelike whistle, "prreep, preep, preep," etc., all in cadence.	Spring peeper
	High, sharp, whistling trill of 4-5 seconds, repeated at regular intervals.	Green toad
	Birdlike whistle of "whit-it-it-it," repeated.	Bird-voiced tree frog
	Soft bass hoot like distant ferry-boat whistle, lasting ⅓ to 1 second.	Colorado River toad
Plain Trills	Repeated clear sharp chirp or trill.	Cliff frog
	Fairly long trill of 5-30 seconds repeated.	American toad
	Shrill trill of 4-12 seconds repeated.	Houston toad
	Low, soft-toned trill of 2-6 seconds, repeated every 20-30 seconds.	Dakota toad
	Quick trill of 2-6 seconds, repeated again in 1-5 seconds, less musical than that of American toad.	Gulf Coast toad
	Very musical, high-pitched trill, 4-10 seconds long, variably repeated.	Red-spotted toad
	Soft but long trill of 10-22 notes, given quickly and at regular intervals.	Yosemite toad
	Trill, 7-10 seconds long, rising in pitch and then stopping quickly.	Southwestern toad
	About 2 low, clear trills per minute, each lasting 1-5 seconds.	Dakota toad
	Steady, repeated, high trill.	Tree cricket

Spring Peeper

Colorado River Toad

Snowy Tree Cricket

* Amphibians generally call from water or moist land. The sounds that frogs and toads make in chorus may also be heard occasionally from individuals. This section includes sounds of insects which resemble those made by birds.

Harsh or Rasping Trills	"Rrrank-rrrank," quickly repeated 20-30 times	Spotted chorus frog
	Rasplike trill, less than a second long, repeated 12-20 times.	Bromley's chorus frog
Ratchetlike Trills	Metallic, vibrant trill, like stroking comb with fingernail, heard after rains.	Western spadefoot toad
	Very noisy, very high, piercing metallic trills, each 15-50 seconds long, with 13-15 seconds between.	Great Plains toad
	Slightly musical, evenly repeated trills, with 7-10 beats to each trill.	Southern chorus frog
Two-part Trills	Somewhat musical, twanging "creek-ek, creek-ek," with last syllable rising, lasting about 1 second; evenly repeated 50-60 times a minute.	Pacific tree frog
Bass Trills	Loud trill, often with chuckling notes, sometimes sounding like pig herd at feeding time.	Crayfish frog
Bell-like or Metallic Calls	Ringing, bell-like, nasal call of "greenk-greenk-greeenk," etc., up to 50-75 times a minute.	Green tree frog
	Single bell-like clear note rapidly repeated, sometimes like rusty pulley wheel being turned quickly.	Strecker's chorus frog
	Series of high, very rapid peeps, like hammer striking steel bar.	Ornate chorus frog
Twanging Calls	Explosive deep twang like "ctang," repeated rapidly 3-4 times.	Bronze frog (also known as green frog)
Chirps or Cheeps	Chicklike "peep-peep-peep," but shrill and ear-splitting in chorus.	Oak toad
	Gentle chirruping, repeated.	Western toad
Loud, Explosive Trills	"Tyeep" and "tut-tut-tut," sung in short phrases with clear, carolling, rising and falling notes; heard just before dawn or in late evening.	Robins
	Trills about ½ second long, each following very quickly, sounding almost like rivetting machine.	Texas toad
	Resonant trill, almost flutelike in tone, 1-3 seconds in length, repeated.	Gray tree frog
Strident Calls	Strident but musical "treet-treet-treet," "cree-cree-cree," or "gru-gru-gru."	Various field crickets
	Insectlike tinkle of "sel-see, sel-see," so shrill as to be beyond hearing of some people.	Little grass frog

Chorus Frog

Spadefoot Toad

Tree Frog

Field Cricket

NONMUSICAL AMPHIBIAN CHORUSES

	Description of Sound	*Source of Sound*
Barking	Explosive "loonk, loonk, loonk," repeated every 1-3 seconds, sounding like guttural "shirrr" at closer range.	Barking tree frog
	Sound like dog barking at intervals of 2-3 seconds, changing to guttural whirring at closer range.	Barking frog

Barking Frog

Narrow-mouthed Toad

Cascades Frog

Wood Frog

Bullfrog

Red-legged Frog

Spadefoot Toad

Tree Frog

Sound	Description	Species
Bleating	Lamblike bleating, sometimes sounding like electric buzzer.	Narrow-mouth toad
	Sheeplike bleat, an explosive and wheezy "aaaah" that lasts 1-3 seconds.	Woodhouse's toad
	Plaintive, anxious bleat like "yee-oow!", descending in pitch.	Couch's spadefoot toad
Buzzing	Buzzing but lamblike bleat, often with early peep, lasting 1-2 seconds.	Eastern narrow-mouthed toad
	Sharp peep followed by angry "bzzzzzz," lasting 2-3 seconds, repeated.	Great Plains narrow-mouthed toad
Chuckling	Soft, chucklelike, grating notes, 4-5 per second.	Cascades frog
Clicking	Rapidly repeated, insectlike "gik-gik-gik-gik," like small stones being clicked together more and more rapidly.	Northern cricket frog
	Sound similar to above, but more metallic.	Southern cricket frog
Guttural Creaking, Quacking, Clacking, and Croaking	Soft, hoarse quacking or clacking.	Wood frog
	Quick, sharp "kak," lasting about ¼ second, repeated.	Plains spadefoot toad
	Deep bass resonant croaking of "ronk, ronk, ronk," repeated.	Bullfrog
	Nasal quack combined with harsh trill, with notes repeated at rate of 80-120 times a minute.	Squirrel tree frog
	Soft, short quack of ¼-½ second, repeated frequently.	California tree frog
Grunting or Growling	Even-pitched stuttering notes like "u-u-u-u-uu-rewrrrrr!", ending in growl, sometimes with soft chuckles.	Red-legged frog
	Quick, explosive, low-pitched grunt, soon repeated.	Eastern spadefoot toad
	Repeated guttural grunting like herd of pigs, sometimes increasing to roar with large numbers in chorus.	Pig frog
	Explosive growling grunt, repeated.	River frog
Hammering or Riveting	Burring rapid bass series of notes, like "tut-ut-tut-tut" or series of hammer blows.	Mink frog
	"Pa-dunk, padunk" like spaced hitting of nails by 2 carpenters in harmony.	Carpenter frog
	Machinelike chorus resembling riveting or a snoring series of very rapid code-like dots and dashes.	Pine woods tree frog
Metallic or Harsh Grating Sounds	Soft but harsh and metallic "cluk," lasting about ¼ second, at rate of 1-3 per second.	Arizona tree frog
	Grating, guttural croak of ⅖-¾ seconds, 4-5 in rapid series, followed by rattle.	Foothill yellow-legged frog
Nasal Sounds	Rasping, high-pitched note, repeated very rapidly, like squeaking of fast-turning rusty wagonwheel.	Mountain chorus frog
	"Konk-konk-konk," repeated about 70-75 times a minute on warm nights, more slowly on cool nights.	Pine barrens tree frog

Leopard Frog

Katydid

Pickerel Frog

Gopher Frog

Rattling	Sound like shaking of wooden rattle.	American toad
Snoring	Rhythmic, rattling snore, repeated.	Leopard frog
	Low-pitched, even snore of 1-2½ seconds.	Pickerel frog
	Bass roaring snore, with chorus, like waves beating on beach.	Gopher frog
Whirring	Whirring, explosive sound at one pitch, lasting 1-3 seconds.	Canyon tree frog
Other Sounds	Evenly repeated, vibrant "rrreek-rrreeek" or "prrreeep," increasing in speed and pitch toward end.	Common chorus frog
	Hoarse "waa-waa" in series of quick rapid calls, 3-5 per second.	Great Basin spadefoot toad
	Sawlike sound.	Spotted chorus frog

Chorus Frog

MUSICAL INSECT CHORUSES

Description of Sound	*Source of Sound*
Repeated cry of "katy-did, katy-did, katy-did."	Angular-winged katydid
Repeated shrill "treee-treee-treee."	Tree cricket

Snowy Tree Cricket

CHAPTER TEN

Smelling Out
Nighttime Odors

BOTH PLEASANT AND UNPLEASANT odors are carried on the night wind, but each can be a help in understanding the outdoors at night. Of course man's nose is much poorer at detecting smells than the noses of most animals, and noses of humans themselves vary greatly in sensitivity, some being almost totally useless because of the constant heavy smells of civilization to which they have been subjected.

Techniques for Sharpening the Sense of Smell

One trick that most woodsmen know to help increase smelling ability is to wet your nose. Also, it is better to go outside on a damp night. On such nights scents are likely to come more strongly to you than on a dry night, since the particles of moisture in the air help carry the smells farther. Another trick is to put your nose low to the ground, particularly if a night breeze is blowing, as the latter usually picks up the strongest scents close to the earth. Even with these tricks, trying to learn about the outdoors in the dark by smelling is more difficult than learning by sight or sound. Hence it is always best to try to tie in the smell you get with a sight or sound of the living or dead things that send it out. Remember noses vary. What may be pleasant to me, may not be to you!

GUIDE TO SOME COMMON NIGHTTIME ODORS
SWEET

Description of Odor	*Source of Odor*
Strong or cloyingly sweet	Honeysuckles or wild lilac in bloom
Soft, warm, and strongly sweet	Water lilies *(Nymphaea)*
Distinctly roselike	Roses in bloom
Rich and tropical	Night-blooming cereus
Fine and fragrant	Night-flowering catchfly
Aromatic	Spicebush *(Lindera benzoin)*, common sagebrush, and eucalyptus
Minty	Wild mint
Warm and sweet	Evening primrose
Alfalfalike	Alfalfa *(Alfalfa lucerne)*, especially new-cut hay
Like fragrance of citrus fruit, faintly sweet	Lemon tree and orange tree
Faint, sweet, violetlike	Violets
Pleasantly vanillalike	Common garden heliotrope *(Heliotropium arborescens)*

Honeysuckle

Wild Rose

Night-blooming Cereus

PLEASANT BUT NOT SWEET OR MUSKY

Description of Odor	*Source of Odor*
Like onions	Wild onion
Like bay	California bay tree (also known as Oregon myrtle)
Like pine needles	Pine trees
Musty	Decaying leaves or mushrooms
Like creosote	Creosote bush *(Larria divericata)*
Rich and earthy	The ground after a warm rain, a sign of good growing things

Evening Primrose

California Bay Tree

Ponderosa Pine

Creosote Bush

Violet

Wild Boar

Swampy, like decaying plants in water	Swamp or bog
Salty or sealike	Salt marsh or estuary
Piglike	Wild boars and wild pigs
Very tangy	Balsam fir
Sulphurous	Sulphur springs

MUSKY

Balsam Fir

Coyote

Description of Odor	*Source of Odor*
Faintly or pleasantly musky and doggy	Coyote, dog, fox, and otter
Faintly musky, pleasant to some people	Skunk scent whiffed from a distance
Faintly or pleasantly musky	Mule and whitetail deer
Strongly but not unpleasantly musky	Beaver, raccoon, bear, muskrat, marten, brown snake, and wood rat
Very strong to nauseating	Skunks, mink, water snakes, and garter snakes
Strong and unpleasant but not usually nauseating	Peccary, weasels, and skunkweed
Acrid	Badger
Strongly musky, coming from dead leaves	Usually a shrew

Striped Skunk

Badger

Least Shrew

Mule Deer

Beaver

Mink

Peccary

Great Horned Owl

Common Crow

Wood Rat

White-footed Mouse

UNPLEASANT BUT NOT MUSKY

Opossum

Cockroach

Description of Odor	Source of Odor
Decaying animal	Nests of owls, eagles, hawks, ravens, vultures, etc.; dens of foxes, coyotes, weasels, martens, etc.; forest or desert floor where dead animals or birds are buried by sexton beetles
Dung	Nests of birds or animals that are not good housekeepers, such as crows, ravens, cowbirds, starlings, wood rats, and white-footed mice
Mousy	House mouse
Strange and fetid	Opossum
Unpleasant and acrid	Cockroach

CHAPTER ELEVEN

Perfecting a
Sense of Direction

Everybody has some direction sense. A person uses it whenever he comes in or out of his house or walks down the street to visit a friend. He has formed patterns of direction and movement in his brain so he can walk to the right place, often even with his eyes blindfolded. This direction sense is probably located in those brain cells that are directly connected with the eyes and ears, but particularly the ears, for it is in the ears that the tiny bones that affect balance and possibly also direction sense are located.

Animals use the direction sense with great efficiency to aid in finding their homes, and particularly to avoid enemies. For example, a rabbit may be seen by night or day dodging with great rapidity through thorny brush and popping into its hole. It is very unlikely that sight alone would allow it to do this, as is shown by the fact that the same rabbit put in an unfamiliar locality will bump into things in its hurry to get away or often become confused. But the rabbit in the brush is going where he has gone many times before; so he has built up a habit that involves his direction sense and can move exactly in the right way and with perfect timing to escape from a dog, cat, or fox that might be trying to catch him.

Setting Up Patterns of Movement in the Brain

You can use this direction sense to set up patterns of movement in your brain that will make it possible to move about in the darkness with much greater ease than other people can. To do this, wander around an area of countryside by day

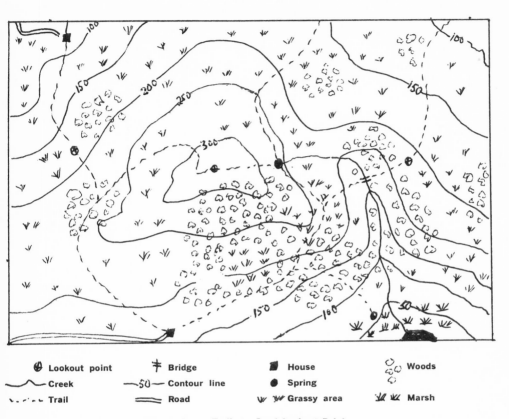

⊕ Lookout point	⟊ Bridge	▰ House	◌◌ Woods
⌒ Creek	—50— Contour line	● Spring	
·.·‛·. Trail	══ Road	ⱱ ⱳ Grassy area	ⱽⱽ Marsh

How to Learn Trails to Good Lookout Points

Go over the trails carefully by day and pace off distances to trail forks, bridges, springs, other landmarks, and lookout points until you can pace off the same distances at night. This will help you to move quietly and keep track of your location with little or no use of a flashlight.

and determine good spots to make night observations, such as the edge of a clearing, a streambank with a hiding place on its top among bushes which is a good vantage point for watching up and down the stream, or a large rock where you can sit and watch up and down a hillside and see creatures not visible from lower down. Another good idea is to look for a hill where you can cross over the top but be hidden by brush or trees, and can suddenly come to an open place on the farther side where animal tracks indicate night activity.

Pathfinding Practice

Going over the trails or routes to these places several times by daylight builds a direction sense which can be used for traveling at night. Repeat the same thing at night until you can find the way from one place to another without difficulty. The darkness will have become quite a familiar place, and you will be able to travel silently from one place to the other by being careful and so possibly come upon animals without their noticing you, especially if clothes and face have been camouflaged to merge with the night and body has been covered with the strong smell of a native plant. The accompanying map shows how this may all be planned. See also Chapter 14 for other suggestions.

CHAPTER TWELVE

Developing a Sixth Sense Outdoors

NEARLY ALL HUMAN beings have at least some of the sixth sense. It is a sense for which we have no definite explanation yet, but it is obvious some people have it quite strongly. I know a former police officer who could tell when a car was coming around the next corner of the street as his own car was driving toward that corner. Invariably there would be one if he said there was. This ability obviously would be of great value to anyone traveling on a twisty road at high speed! Many more people have the ability to tell when somebody is staring at them, even though the stare is coming from behind them. I have felt this feeling many times and looked up or around to see somebody watching me.

Uses of the Sixth Sense

In the outdoors the sixth sense can often help the wildlife observer to know when he is being watched by a wild animal or bird and to locate the watcher. It can even warn him of danger if the creature is dangerous. At all times it helps him to be more aware of his surroundings and may lead to discoveries that he would otherwise miss.

How can the sixth sense be developed? Usually by practice, like anything else. When you are walking down a street, in a park or in the countryside, settle your body into a feeling of balance by thinking of its center, which is about two inches below the navel, as a focal point from which all your actions and thoughts flow out like rivers of power. To this point all the surroundings flow in through the senses of touch, seeing, hearing, smelling, tasting, and the sixth sense. Be dynamically relaxed while walking so that tenseness caused by fear, worry, discomfort, or other distracting feeling is eliminated. Grow in a dynamic feeling of awareness of everything in the vicinity. Even if the sixth sense does not seem to develop, your other senses, so necessary for successful observation, will be sharpened. Gradually, you will be able to sense things with the sixth sense, a faculty which will prove surprisingly helpful.

Exercises for Developing the Sixth Sense

Here are some exercises to increase the sixth sense, with the help of a partner. Trade off with him in your actions so he also can increase his sixth sense. Do each testing with a feeling of relaxation, getting rid of all worry if possible. This is very important.

1. Walk along slowly with your right or left arm held out in a relaxed manner. Have your partner approach you silently from behind but at a little faster speed and reach out to grab your arm. You will be looking straight ahead, but at the moment he is about to touch your arm, you will feel this before he actually does so. At that moment move your arm swiftly forward so the grab does not reach it. Consistent ability to do this just at the right moment and without being touched is evidence of a developing sixth sense.

2. Have a partner stand about fifty feet away, looking up at some distant object, while your back is turned to him. Tell him to wait a bit and then suddenly turn and stare at you very hard, but without making any noise. If you feel that he is staring at you, turn and look at him. If the sixth sense is developing properly, you will be able to turn every time just a moment or so after he begins to stare. Avoid nervousness, as it will cause you to turn too soon.

3. Have your partner blindfold you. Then walk toward a wall or another human being with arms outstretched, very slowly. When you feel that wall or that person is only an inch away, stop. If you can do this consistently, you are developing the sixth sense.

The sixth sense can help in many ways at night if it becomes strong enough. You can tell, for example, the right direction to go to reach a trail or road when there is danger of becoming lost. And you may be able to sense where a particular animal or bird, seen a few moments before, has gone, and so follow it to see it again. This sort of sensing increases with greater relaxation and a heightened awareness of the surroundings.

Importance of Feeling Akin to All Life

I feel strongly that at least a part of the sixth sense grows out of an increasing sense of kinship with all life. Most people look at living creatures and plants as things with no particular feeling for them. But when you realize that all living things have much in common with you and came from the same source of all life, then it becomes easier to turn toward them with a feeling of love and understanding. They too can be hurt; they too have feelings for other living things, particularly their mates, and they can quite literally feel a human being's love as any truly expert trainer or keeper of animals, wild or tame, can testify. Any person fond of dogs, cats, or horses knows how they respond to affection and kindness by seeking to be near him, and at least in the case of dogs, being eager to do what he wants them to do. By extending the same affection and understanding to other living things, the night-time explorer causes them to feel an influence of kinship from him, and, as a result, animals may not only show themselves to him more often but even subconsciously help him to locate their presence through the sixth sense. I cannot prove this, of course, but a lifetime of being a naturalist has given me a very strong feeling that this is so. Several times when I have been sitting very quietly in some wild place, particularly at night, a wild animal or bird has come out of the silence to visit me. Whether it knew I was there or not, it came because it felt no vibrations of danger from me, only great curiosity and love.

CHAPTER THIRTEEN

Staying Clear
of Trouble

IT IS BETTER to accent the positive instead of the negative at night, especially because fears of the dark tend to magnify what are often comparatively small dangers to something big. And fear itself, with its danger of making one run or exhaust oneself when such action is not at all needed and so stumble into situations which could be dangerous, usually vanishes when one understands the actual realities of the night and uses common sense to avoid or minimize hazards.

Of course you can avoid almost all dangers by watching at night from a screened cabin porch or similar safe place on the edge of the out-of-doors or even in the midst of it and never getting caught in the dark. Many interesting things can be seen from a porch, and the beginner would do well to start his night observations in such a way. For those who are either a bit more adventurous or who are caused by work or accident to move through the darkness in open country, the following pointers may be helpful.

Safety Pointers

1. Keep cool at all times. Remember that most dangers in the dark are either imagined or greatly exaggerated. Remember that most forests, deserts, and so on are far less dangerous than a highway or even many a city street.

2. Bring along any of the tools mentioned below under Tools to Aid in Safety that would be particularly helpful where you are going. Being well equipped is good preparation for potential crises and generates self-confidence.

3. Avoid dangerous areas, such as swamps, rivers with quicksand, areas where many poisonous snakes are reported, and so on. You can usually inquire at a local store, particularly where guns and other equipment is sold to hunters, about local conditions and hazards.

4. Poisonous snakes want to avoid you just as much as you wish to avoid them. By avoiding moving too quickly in brushy, rocky, or swampy areas, you will not surprise a snake so suddenly that it will strike before escape is possible. Watch where you step and where you put your hands at all times, and there will be little likelihood of trouble.

5. Standing still and speaking softly calms most animals; a light flashed in the eyes causes most animals to retreat.

6. If you go alone or with a small party on a trip into the night, leave word with a responsible person telling the destination of the trip and the expected time of return. This makes it easier for others to rescue you in case of a mishap.

7. Poison sumac, oak, and ivy are illustrated here. A flashlight will help you identify these so they can be avoided.

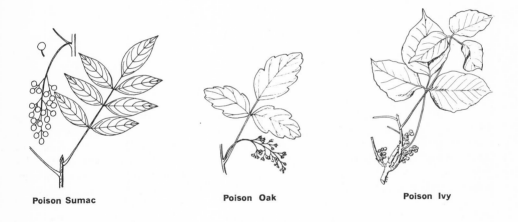

Poison Sumac   **Poison Oak**   **Poison Ivy**

8. If you are going on to someone's property, be sure to get the owner's permission and let him know when you are going and where. This will avoid sudden confrontation with an irate landowner who might even use force to ensure your leaving promptly.

9. If you do become lost, don't panic and run, but stay calmly in one place until daylight. If matches are available, build a fire, but be careful to clear a space of at least eight feet around the fire if possible, so that it will not spread. (Of course, in seasonal fire-hazardous areas, especially in the drier chaparral and forest areas, it would not be permitted to build a fire.) A map of the area, carried in pack or pocket, will be of great aid in getting you back to home base.

10. It is best to move into wild country at night in very small bits at first until you have gained the experience and confidence to go farther.

Tools to Aid in Safety

A flashlight. Used with a red plastic cover, it can be turned on wildlife at night without bothering it. With the plastic cover off, it can help you to reach safety and avoid dangers while traveling from one place to another. A bright light flashed in the face of a savage animal while shouting in a commanding voice often drives it away.

A good knife. Useful in the following ways: cutting a pair of sticks and some willow withes to make a splint for a broken leg or arm; cutting lengthwise (never crosswise) incisions on an arm or leg for poisonous snakebite to get out both blood and poison; making a crude fishing pole or snare if lost in a wilderness area and needing food; shredding bark to help start a warming, cooking, or signal fire. If used for cutting around a bite, it should be sterilized with a burning match.

A compass. Useful in finding one's way when lost, especially with the aid of a map. Since most people walk in circles when lost, the compass enables the user to walk in a straight line and so eventually get to some safe area.

Matches. Useful in sterilizing a needle or knifeblade whenever an incision in the skin or flesh is needed for snakebite treatment or to remove a big sliver. Can be used to start a signal fire and to help you stay warm when lost. Should be used with great care by clearing space down to mineral soil at least eight feet around the spot where fire is started. Fire should be completely put out with water or plenty of dirt when finished. If the woods are wet, find dry bark and wood under a fallen log, and shred with knife to make fine dry material to start fire.

First aid and snakebite kits. Should be taken on night excursions of any distance when in poisonous snake country and in warmer parts of the year. Carefully study directions for use, usually present in each kit. First aid kit should be along on any lengthy hike any time of the year. An elastic-type bandage for sprains is an excellent aid.

Belt hatchet or machete. Useful in cutting through brush, building shelter, making fire, and possibly for defense.

Fishing line and pair or so of hooks. Useful in emergency to get food from stream or lake. Small sinkers would be useful too, but they can be made from rocks. The line can also be used for making an animal snare for catching small animals for food in emergency, as shown.

Animal Snare Made with Fishing Line

Boots. Better protection against snakebites, thorns, stickers, and so forth than shoes. Should be well oiled so they can be used for walking through wet places. Should be no more than ankle-high.

Map. Should cover area in which you are walking, to show you points of reference and how to get out to civilization.

Avoiding Fear and Recklessness

Fear may cause a person to run wildly and so land, without meaning to, in a place of greater danger. Recklessness may do the same thing for opposite reasons. The night obviously does have dangers such as unexpected cliffs, holes, slimy or moss-coated rocks in streams, swamps, thick clusters of trees or brush, and sharp dead jutting branches, among which it is possible to get injured, entangled, or lost. Also, in dry streambeds there is the chance of flash floods. One should honor trespassing signs, and beware of going into pastures where there are cattle. Cows with calves, frisky young range cattle, and especially bulls can be perilous. Also in some localities wild boars can be a danger. So a wise night explorer carefully avoids both fear and recklessness, thus safeguarding his person, especially his eyes at all times. This is why it is best for beginning explorers of the night to move into the darkness only short distances at first to get a feel and understanding of the outdoor night and to develop a direction sense as outlined in Chapter 12.

Fear is largely caused by lack of knowledge. Studying this book carefully and proceeding slowly at first when getting out into the night will get rid of most fear. But fear may be a useful emotion in case of real danger as long as it does not give way to panic, for it may prevent people from doing something foolish. Deep, slow breathing and a realization that most dangers can be avoided by holding still and being calm will help to combat fear.

Recklessness can be caused by overconfidence or even a desire to show off. The night in the outdoors warrants a healthy respect. If you grant it this respect, it will pay ample dividends.

CHAPTER FOURTEEN

Learning to Watch
and Attract Wildlife

WISDOM OFTEN COMES best out of silence. To be silent in the heart of the outdoors and to listen with heart and mind is to do what the great sages and saints of the past have done, and always with benefit. The talk of many people shows a fear of the silence; they are afraid to be alone and to listen. But the wise man draws from the silence knowledge and strength, companionship with all life, and a peace of the soul that is strangely invigorating. So did such men as Thoreau, Seton, and Audubon find strength in the silence.

Wisdom of the outdoors greatly increases from night observations. Like life anywhere, but much more so because of the mystery the darkness brings, life after dark can be likened to a giant puzzle whose interlocking parts you begin gradually to put together. Unlike most man-made puzzles, the curiosity, wonderment, excitement, and adventure can last forever, since there are always new mysteries. Here we can mention only small parts of the great puzzles as examples of what curiosity, exploration, and growth in wisdom of the dark can begin to unravel.

Watching Animal Trails

During wandering in the outdoors by day you may begin to notice trails left by various creatures, from the tiny passageways of mice and shrews through the grass stems up to much larger marked trails through the woods or brush made by elk, moose, and deer. Much of the activity along these trails is likely to be at night when the animals feel they are least observed. Yet approaching these trails in the dark will cause noise which is likely to drive away all nearby wild creatures so that you will have little chance to see them. Growing outdoors wisdom makes it plain that these animals can best be seen by stationing yourself near one of these trails at a good observation point, being very quiet, and emitting no betraying scent. In fact, it is possible to build a blind at such a place so that you will truly be invisible to the animals. Such a blind is described and pictured in Chapter 1. Cover your clothes and body with the strong-smelling scent of a plant like balsam, sagebrush, or honeysuckle to disguise the human scent and sit quietly after dusk until the animals come. Sometimes they come soon, other times after several nights, but patience brings a reward.

Another way to use trails of animals by day is to mark a place where two or more parts of the trail can be seen at once. The accompanying map shows such a vantage point.

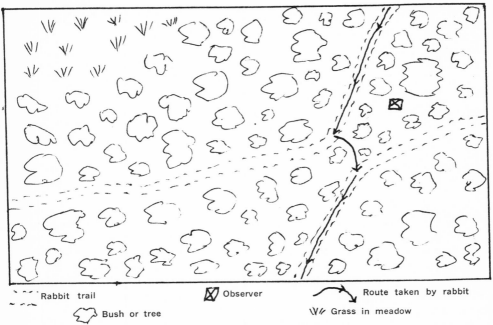

- - - Rabbit trail   ⊠ Observer   Route taken by rabbit

🔾 Bush or tree   \W/ Grass in meadow

Vantage Point for Watching Rabbit Trail
Note how rabbit changes trails to foil pursuers.

Rabbits and hares have a trick of running along a trail at night and then suddenly leaping sideways at right angles so as to land on a different trail down which they run hidden by bushes. Often this happens just after they have made a turn in the first trail (see map); so when a coyote, fox, or dog is chasing them, their sudden leap to the side just when out of sight will seem to make them disappear into thin air. You can find such places where rabbits or hares do this by exploring the area by day and looking for tracks. Since they are likely to execute such a maneuver in a favorable location over and over, the place where the feet land after the long jump may be deeply marked. Use your wild wisdom to help find a vantage point for watching this action at night, especially on a moonlit night. Or, if you see a rabbit or hare do this disappearing act when you disturb or chase it, use a flashlight to discover where and how the trick was done.

Knowing that such a trick may be about to happen will help you turn your eyes to the place where the rabbit or hare is likely to be seen next. Often an animal wants to watch you more closely after it has run off; so it circles to left or right and watches from behind a bush or rock to see what you are up to. A part of growing outdoors wisdom is the ability to put yourself inside the mind of the animal and figure what he will do, so that it becomes easy to turn, look, and see where it is watching you.

Mice and rats give evidence of their presence and their activity by leaving their droppings near holes or trails where they are most active. If your flashlight uncovers such fresh droppings, or you discover them by daylight, put a camp chair or other comfortable seat near a place, cover your clothes with strong-smelling plant scent, and sit quietly to watch for the activities of these animals in the moonlight or by the light of the red-covered flashlight.

As Chapter 11 explained, direction sense in the dark is increased by getting familiar with a particular territory so you automatically move in the right direction. In the same way, going over and over a particular acre or two of ground by day develops familiarity with all the landmarks—the big rocks, trees, important bushes and so on, and especially all the animal trails. Come back at night and learn again how to find these things so that you can move quietly from place to place and never get lost. Then it becomes possible to watch the wild creatures and birds and know best where to expect them because you have seen their signs before.

Another wisdom that comes through experience in the dark is the knowledge of when and when not to move. If a coyote, fox, or wildcat is near, it is best to stand or sit absolutely still as these creatures are extremely quick to catch sight of any strange movement. Not knowing what it is, but being suspicious, they may disappear in a flash. Animals like rats, mice, rabbits, and deer notice movements, but not so sharply; and it may be possible to move very quietly toward them, provided you do not smell like a human and are camouflaged by clothing and a smudged face that blends with the dark. Watch carefully and move just when the animal is feeding with head down. While eating, its eyes and ears are temporarily not as alert as when its head is up. It is necessary to learn patience and not be in a hurry, but to move only when the time is right.

Techniques for Attracting Wildlife

Many birds and animals are curious when they hear, see, or smell something unusual. As soon as they recognize a human, they try to disappear; so the trick is to be disguised while you use something else to produce the desired effect. For example, hang tiny mirrors or bright pieces of metal in a tree by day. Then come back at night, assume a comfortable position, and watch for what animals or birds are attracted by these things when the moonbeams flash on them. Wood rats in particular are nearly always attracted by bright things and will go to great lengths to try to get them for trade, leaving a rock or stick in the place of what they have borrowed.

The ability to copy fairly exactly animal or bird sounds will often attract different creatures. Holding a flat blade of grass against your lips and blowing, when done after sufficient practice, produces a squeaking sound like a mouse in the grass. This sound can be made to attract a fox, coyote, wildcat, or other predator to come near. Certain companies produce special whistles that duplicate fairly exactly a squeak or scream of a rabbit in peril or pain. Such sounds are almost irresistible to a neighboring fox, great horned owl, wildcat, and so forth. Have your head protected, though, by a good cap, so an owl will not strike and hurt you.

With your voice you can often, after practice, imitate the hoots or other cries of owls and thus attract these interesting creatures. The whippoorwill and the poorwill are also attracted by imitations of their calls. Along the beaches and shores various shore birds often cry in the dusk of evening or dawn, and you can copy these cries while standing perfectly still until the birds come close.

Another stratagem is to go to a zoo and record on a portable tape recorder the sounds of various zoo animals and birds that are also found in the countryside. Record as many sounds as possible of each creature, using a homemade large "ear" made from aluminum foil and wired to the microphone (see diagram) so as to

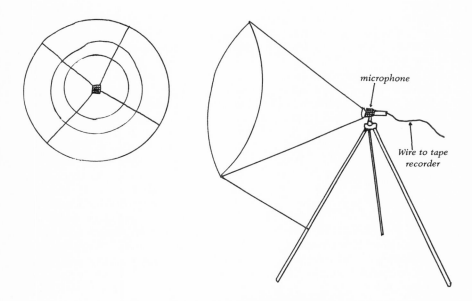

"Ear" for Use with Tape Recorder
Left: **Wire frame to hold aluminum foil cone.** Right: **Aluminum foil cone wired to microphone of tape recorder and to tripod support.** (Note: Cone can be made much shallower if this brings in sounds better. Do some experimenting to see what is best.)

facilitate catching the sounds loud and clear on the recorder. Then take the recorder into the countryside at night and play various sounds to attract the wild creatures to your hiding place. Recordings of the mewing of kittens, the squeaking of pet mice or rats, and the whine of small puppies may also bring other creatures near to see what is up.

Other animals can be attracted by baits. Fresh meat put on a rock or log near where ring-tailed cats or raccoons have been seen or heard may soon attract them to the food in the night. Fresh greens attract rabbits or hares, particularly if the surrounding area is without such fresh green things because of lack of rainfall. Even on a dark night you can catch glimpses of these creatures coming to the food by using the always reliable red-covered flashlight.

There are many other such tricks that can be used to attract and watch wild creatures if you keep on learning and do some thinking about new ideas. Good luck!

CHAPTER FIFTEEN

Using the Dark
to Become Nature-wise

THIS BOOK CANNOT close without suggestions for particular ways to enjoy yourself in the dark and learn at the same time. All such projects involve a fundamental rule, especially now that we are pollution- and ecology-minded, which is to do nothing that is harmful or annoying to other human beings and to respect all animal and plant life to insure this earth of ours maintains its beauty and interest for future generations.

Testing Animals' Reactions to Strange Noises

It is great fun to see how a wild animal or bird will react to a noise that is strange to it. Make use of a portable tape recorder for this experiment, making recordings, for example, of strange animals and birds that are found in a zoo. Often, when the native birds and animals hear these sounds, they become very curious and come near to see what is causing the disturbance, never suspecting that a human is near watching them. Some normal sounds can be made to sound strange by running them on the tape recorder at slow or fast speeds. Also, scraping wood together, shaking seeds in a rattle, and other mechanical methods produce unusual noises that will attract the curiosity of wild creatures.

This kind of experiment may not have been tried in some areas, and you should keep careful written records of just how the animals and birds react and exactly what they do. Some creatures may even come close to attack the maker of the strange sound. If they are large, it may be necessary to notify them that a human being is present; but smaller creatures should be very amusing to watch.

Testing Animals' Reactions to
Strange Objects or Changed Conditions

Cardboard copies of what look like owls, using fluorescent paint to make the eyes shine in the moonlight, can be placed in strategic places in trees to see how these affect the activities of rabbits and other creatures in a forest clearing or meadow. A rubber copy of a large snake can be placed across a trail frequented by rabbits, wood rats, or jumping mice and rats to see how they react when they see it. This will give you insight into the defense mechanisms or reactions of different animals and birds.

A rubber or other type model of a mouse can be put on top of a stump or post or in an open space on the ground. If it can be moved some way with a tiny invisible wire or thread so that it looks realistic, it will attract an owl and you can observe not only how the owl attacks but how he reacts when he finds he has caught something different from what he expected. Sometimes these reactions are very amusing. To make the mouse move in two directions, use two wires or strings, one to pull it over to one side of the stump, and the other to pull it back to the opposite side (see diagram). Models of other small animals can be used in the same way in other situations to attract foxes or wildcats, etc.

ring on stiff wire strung between two trees

String attached to ring above mouse and to ring near tree

Device for Attracting Owls
Stuffed mouse is attached to stiff wire extending down from ring. Observer pulls strings from either side to make mouse move back and forth on stump.

After fresh snow has fallen, come out on snowshoes and carefully put out baits of fresh meat or carrots and other vegetables in places where foxes and rabbits come. Come back in the dawn and see by the tracks how the wild animals have reacted to this unexpected food. Or go out at night after the food is put out and watch from a comfortable perch. If it is a moonlit night, you should see some interesting sights.

Block of Salt in Cage for Attracting Porcupines

If there are porcupines in the area, lay out a trail of salt down to a clearing in the woods where you have put a block of salt into a barlike box structure (see diagram). Stand guard from your post. Porcupines love salt and will follow this trail. They will try to get into the box to get at the salt. Watch their antics and their reactions to each other. As they are among the most comical of animals, this should be very amusing. After a while try making a noise like a dog barking and see their reactions.

If there is a den of a fox, coyote, or wildcat in the neighborhood, set up a blind near it by daylight, camouflaging it well as described before. Come back in the early evening of a moonlit night, or bring the trusty red flashlight, and wait patiently until the cubs come out to play. Stay absolutely still, and you will see some extremely amusing antics and learn a lot about the play of young wild animals.

Exploring the Mysteries of Night-blooming Plants

Certain plants, such as the evening primrose, the honeysuckle and the night-blooming cereus, open their petals not in the daytime, as do most plants, but in the evening or night. Why they do this, how long they do it, and what determines the length of the bloom is a mystery worth unraveling. I could tell you why now, but it would take away the fun of discovering for yourself. (Also, I could be wrong!) The exploration of this strange phenomenon leads to some extraordinary sights that are very beautiful and exciting to observe.

Find one of these plants in a garden, city park, or the countryside and watch it as evening comes, especially on a warm late spring or early summer day. Watch with extraordinary care and patience, for you are going to behold the magic of seeing a plant moving not in stimulation to the light, but in stimulation to the coming dark, and every moment should be seen to be understood.

After the flower has opened its petals in a most intricate and graceful manner, certain creatures that have tongues especially adapted for getting nectar from such flowers come to the flower. Watch carefully how they feed and, if you can get close enough, see how they carry away not only the nectar from the flower but also some of its pollen, which is then able to fertilize other flowers of the same kind. Watch the flowers and their small visitors through the dark hours and record how long the petals stay open. When do they close and what do you think happens to make them close at the hour they do? It is a mystery worth solving.

Try experimenting with the flowers to see if they react to the time of day and night or entirely to light and darkness. Put a tent or tarp over a group of flowers in the middle of the afternoon so they are in semidarkness and note if this makes them bloom earlier or whether they wait for the correct hour of evening when the other flowers of their kind will bloom. Test them when darkness comes by hanging a brilliant gasoline lantern above them so they stay in bright light. Do they stay in closed form without opening up because of this light or do they open up anyway as the other flowers of their kind are doing?

You will learn from both of these experiments that plants are not so far different from animals in their reactions. As unique living things, they deserve our respect and protection. Try to find other plants that open up in the night, and experiment with them. One such plant is the almost animallike, but very primitive slime fungus, which climbs over wet logs like a giant, moving amoeba and often glows with ghostly light in the darkness.

Grading the Performance of Your Nighttime Senses

Every human being probably reacts in a different way to the darkness of night. Some people have eyes that adapt slowly to the darkness; the eyes of others adapt quickly. Why not have the fun of seeing where you stand? It will take about half an hour. Select either an open field or a woods on a night that has no moonlight, but does have starlight. Before the darkness falls, set out a row of white stakes in the ground, about ten feet apart, all equally white and all put the same distance into the ground with the same amount of white showing aboveground. Be sure you can stand in one spot in the daylight and see all of these stakes stretching away a distance of at least 120 feet. When the night is thoroughly dark, take a flashlight and walk to the same place to look at the stakes. Have a watch for timing; turn out the flashlight and see how long it takes before you can spot the farthest stake in the light of the stars, and also how long before each of the stakes becomes visible. Record this.

Eyes which react normally to the darkness are able to see the farthest stake in half an hour. If your eyes are very good in the dark, you will do better than this time, and if poor, it will take longer or the stakes may not become discernible at all. By the way, wear glasses, if you normally do.

To test your hearing of night noises, have a friend take a box of about ten stones, weighing by number from 1 to 10 ounces, and have him drop these stones in a bucket of water 100 feet from you, one at a time in an open field and again in dense trees.

The dropping should be done after a prearranged signal, such as the flashing of a flashlight. Flash the light back when you hear a stone drop. Your friend will reveal whether the first stone dropped was the 3-ounce one, the 5-ounce one, or whatever. This will give you an idea of how good your hearing is in the dark, and a basis for comparison. Sound carries far at night, and the distance may have to be increased.

Team Contests of Nighttime Seeing and Hearing Ability

An exciting game for testing both vision and hearing at night is played by forming two teams of equal numbers of players on each side and placing the teams 100 yards apart in either woods or an open field on a moonlit or starlit night. Each player should be twenty feet from the next player on his team. Each player has a flashlight, which he can use, however, only when notified by a referee who stands by the edge of the field. Each player on a team has a number, so if there are five members of a team, they are numbered from one to five, and the same with the other team. The two teams start to move slowly towards each other, trying to be as noiseless as possible, holding their lights in front of them but no light showing until signaled. When the referee thinks the time is right, he calls, "Number twos flash lights!" or "Number twos and fours flash lights!" When the light is flashed on by each player, he tries to flash it directly at the opposing player, but he must hold the beam steady without moving it. If the beam hits fully on the opposing player, that is one point for his team. If he himself is hit by an opposing team's light, it is one point for the opposing team. As the players creep toward each other, they try to change position after each showing of the flashlight so they will be harder to find. The game ends when every player has been caught at least once in a flashlight's beam. Points are then added up to see which team has the most points. This game teaches alertness, and trains in moving silently, while developing the art of seeing and hearing in the dark.

Besides the above game, try testing the hearing and sight of friends in the same way you were tested above in this chapter.

Still a third game involves forming the same kind of teams at the same distance. The teams move towards each other, but throw fir cones at members of the opposing team whenever one hears a sound or notes a movement. Whenever a player is hit by a fir cone, he must admit it, and this is one point against his team. The game lasts for half an hour and the winning team has the most hits. One team wears a band on the left arm; the other team wears it on the right arm.

Fir cones are very light and can do little damage.

Making Wildlife Sound and Sight Maps

Map an area of one to four acres by daylight so that you know where each tree, shrub, large rock, and other prominent object is found. Set up two good blinds, with comfortable chairs in each, in two good observation points inside the area. Use a compass to draw on the map from these two observation posts lines to each prominent object inside the area that can be seen from each post. The first map in this chapter shows how to do this. Have a friend who is interested take one observation post each night for an hour while you take the other during the same hour.

The project is to map each sound and sight heard or seen each night of watching. The purpose is to establish a pattern in sights and sounds which will help reveal not only the birds and animals, but what relationship they have to each other. The

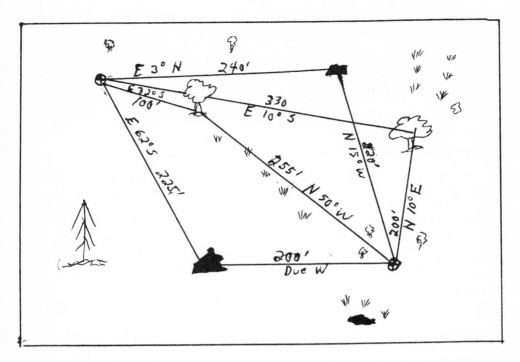

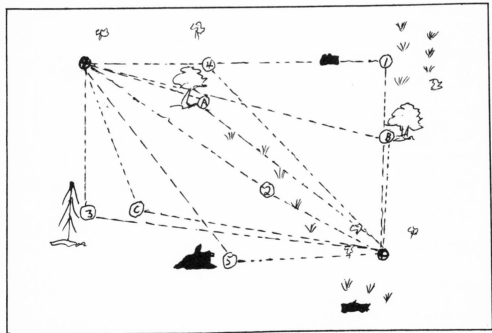

How to Make a Wildlife Sound and Sight Map

Top: **Map of one acre with two observation posts. Lines are drawn from each post to large visible objects in area to give proper distances from observation posts.** Bottom: **Same map, but showing sources of sounds (numbers) or sights (letters), checked by two observers for compass directions and distances from each observation post. Dotted lines indicate cross-checks. Sounds and sights are as follows: 1. Squeak of mouse 2. Rustling sound, probably mouse in grass 3. Hoot of owl 4. Hoot of answering owl 5. Snuffling noise of digging skunk A. Owl dropping on mouse B. Cottontail diving for hole under tree C. Skunk crossing field in moonlight**

second map shown with this chapter illustrates such a pattern. This is a major project, but an extremely interesting one and, ideally, should be done only by two people eager and willing to carry it through for at least four nights in a row.

Taking an Animal Census from Tracks Made at Night

Another project is to study an area, say an acre, where there is snow, mud, or dust in enough quantity to give a good showing of tracks made in the night. Or artificial dust or mud can be spread in an area each evening and smoothed over, then studied again the next dawn to see what the tracks show. A map of the tracks and the stories they tell can be made each day, and all compared at the end of the week to see what patterns of life and activities were formed in the area during the week. From the tracks you can make a crude census of the number and kinds of animals and other creatures that inhabit your study area. This project would work particularly well around a spring in the desert where many animals and birds come to drink, feed, and hunt during the night.

Calling Owls

Two friends can have a lot of fun going through a woods separated by about 100 yards and hooting like the various owls that inhabit that wood. Try to make your voices resemble calling back and forth and imitate the owls as closely as possible. Move very quietly all the while. If the imitations are at all good, some real owls should answer, and if your clothes and face are well camouflaged to look like part of the woods, the same owls may soon come close to investigate.

Other Night Projects of Interest

From a cabin perch in the wild country or surroundings many projects can be carried out. One would be mapping all the area within sight of the cabin by daylight so you can locate each tree, large shrub, and large rock by compass bearing, both on the map and then by sight from the cabin; then using the same landmarks at night to identify the direction of each sound heard and each sight noted during an hour or so of listening and watching each night. After several nights a very interesting pattern of life should begin to form.

Another project would be to try to classify all the sounds heard according to the categories given in Chapter 9 of this book, and by doing so, identify as many creatures as possible by sound alone. Then check this by trying on other nights to get as close to the sounds as possible, either by stalking or by setting out hiding places in the strategic spots to catch sight of the creatures that make the noises.

These suggested night projects are intended mainly to get you to think up projects of your own. So do not be content with carrying out the suggestions in this book. Not only does the night have a profusion of forms of life, but it also affords opportunities for myriads of potential projects to make the hobby of studying that life a great adventure and great fun.

Suggested References

General

Benton, Allen H., and Werner, W. E. *Field Biology and Ecology.* New York: McGraw-Hill Book Co., 1966.

Milne, Lorus J. and Margery J. *The World of Night.* New York: Harper & Row Publishers, 1956.

Palmer, E. Laurence. *Fieldbook of Natural History.* New York: McGraw-Hill Book Co., 1949.

Prince, J. H. *Animals in the Night.* San Francisco: Tri-Ocean, 1968.

Birds

Bent, Arthur C. et al. *Life Histories of North American Birds.* Edited by Arthur C. Bent. Several vols. New York: Dover Publications, 1951.

Brown, Vinson; Weston, Henry J.; and Buzzell, Jerry. *Handbook of California Birds.* 2d rev. ed. Healdsburg, Calif.: Naturegraph Co., 1972.

Forbush, Edward H. *Natural History of American Birds of Eastern and Central North America.* Boston: Houghton Mifflin Co., 1939.

Peterson, Roger Tory. *Field Guide to the Birds.* Rev. ed. Boston: Houghton Mifflin Co., 1947. (Covers eastern land and water birds.)

————. *Field Guide to Western Birds.* 2d rev. ed. Boston: Houghton Mifflin Co., 1961.

Robbins, Chandler S.; Bruun, Bertel; and Zim, Herbert S. *Birds of North America,* A Golden Field Guide, illustrated by Arthur Singer. New York: Golden Press, 1966. Copyright 1966 by Western Publishing Company, Inc.

Zim, Herbert S., and Gabrielson, Ira N. *Birds,* A Golden Nature Guide, illustrated by James Gordon Irving. New York: Golden Press, 1956. Copyright 1949, 1956 by Western Publishing Company, Inc.

Fish (*see also* Freshwater Life *and* Seashore Life)

Eddy, Samuel. *How to Know the Freshwater Fishes.* Dubuque, Iowa: William C. Brown Co., 1957.

Freshwater Life (*see also* Fish, Mammals, *and* Reptiles and Amphibians)

Coker, Robert E. *Streams, Lakes, Ponds.* Chapel Hill: University of North Carolina Press, 1954.

Morgan, Ann. *Field Book of Ponds and Streams.* New York: G. P. Putnam's Sons, 1930.

Pennak, R. W. *Fresh-Water Invertebrates of the United States.* New York: Ronald Press Co., 1953.

Russell, Franklin. *Watchers at the Pond.* New York: Alfred A. Knopf, 1961.

Usinger, Robert. *Life of Rivers and Streams.* New York: McGraw-Hill Book Co., 1967.

Insects and Their Relatives (*see also* Freshwater Life *and* Seashore Life)

Borror, Donald J., and White, Richard E. *Field Guide to the Insects of America North of Mexico.* Boston: Houghton Mifflin Co., 1970.

Buchsbaum, Ralph. *Animals Without Backbones.* Rev. ed. Chicago: University of Chicago Press, 1948.

Larson, Mervin W. and Peggy P. *Lives of Social Insects.* New York: World Publishing Co., 1968.

Mammals (*see also* Freshwater Life)

Burt, William H., and Grossenheider, Richard P. *A Field Guide to the Mammals.* 2d ed. Boston: Houghton Mifflin Co., 1964.

Caras, Roger. *North American Mammals.* New York: Hawthorn Books, 1967.

Palmer, Ralph S. *Mammal Guide: Mammals of North America, North of Mexico.* New York: Doubleday & Co., 1954.

Pocket Guide to Animal Tracks. Harrisburg, Pa.: Stackpole Co., 1968.

Plants

Abrams, Leroy. *Illustrated Flora of the Pacific States.* 4 vols. Stanford, Calif.: Stanford University Press, 1921-63.

Britton, Nathaniel, and Brown, Addison. *Illustrated Flora of the Northern United States, Canada and the British Possessions.* New York: Dover Publications, 1970.

Cronquist, Arthur; Holmgren, Arthur; Holmgren, Noel H.; and Reveal, James L. *Flora of the Intermountain West,* vol. 1. Darien, Conn.: Hafner Publishing Co., 1970.

Grimm, William Carey. *The Book of Trees.* Harrisburg, Pa.: Stackpole Co., 1962.

————. *Home Guide to Trees, Shrubs, and Wildflowers.* Harrisburg, Pa.: Stackpole Co., 1970.

——. *Recognizing Flowering Wild Plants*. Harrisburg, Pa.: Stackpole Co., 1968.

——. *Recognizing Native Shrubs*. Harrisburg, Pa.: Stackpole Co., 1966.

Jaeger, Edmund C. *Desert Wildflowers*. Rev. ed. Stanford, Calif.: Stanford University Press, 1968.

Montgomery, Frederick H. *Native Wild Plants of the Northeastern United States and Canada*. New York: Frederick Warne & Co., 1956.

Peattie, Donald C. *A Natural History of Trees of Eastern and Central North America*. Boston: Houghton Mifflin Co., 1966.

——. *A Natural History of Western Trees*. Boston: Houghton Mifflin Co., 1953.

Radford, Albert E. *Manual of the Flora of the Carolinas*. Chapel Hill: University of North Carolina Press, 1968.

Round, F. E. *Introduction to Lower Plants*. New York: Plenum Publishing Corp., 1969.

Reptiles and Amphibians *(see also* Freshwater Life)

Conant, Roger. *A field Guide to Reptiles and Amphibians*. Boston: Houghton Mifflin Co., 1958. (Covers eastern half of the United States and Canada.)

Oliver, James A. *A Natural History of North American Amphibians and Reptiles*. New York: Van Nostrand-Reinhold Books, 1955.

Stebbins, Robert C. *A field Guide to Western Reptiles and Amphibians*. Boston: Houghton Mifflin Co., 1966.

Seashore Life

Braun, Ernest, and Brown, Vinson. *Exploring Pacific Coast Tide Pools*. Healdsburg, Calif.: Naturegraph Co., 1966.

Evans, Idrisyn O. *Observer's Book of the Sea and Seashore*. New York: Frederick Warne & Co., 1962.

Guberlet, Muriel L. *Animals of the Seashore*. 3d ed. Portland, Oreg.: Binfords & Mort, Publishers, 1962. (Covers northwest Pacific coast.)

Hay, John and Farb P. *Atlantic Shore*. Boston: Atlantic Monthly Press and Little, Brown & Co., 1968.

Hedgpeth, Joel. *Common Seashore Life of Southern California*. Healdsburg, Calif.: Naturegraph Co., 1961.

——. *Introduction to Seashore Life of the San Francisco Bay Region and the Coast of Northern California*. Berkeley: University of California Press, 1962.

Hinton, Sam. *Seashore Life of Southern California*. Berkeley: University of California Press, 1962.

Miner, Ralph Waldo. *Field Book of Seashore Life*. New York: G. P. Putnam's Sons, 1950. (Covers the east coast of the United States and Canada.)

Morris, Percy A. *Nature Study at the Seashore: Exploring with Your Camera*. New York: Ronald Press Co., 1962.

Ricketts, Edward F., and Calvin, Jack. *Between Pacific Tides*. 4th ed. Stanford, Calif.: Stanford University Press, 1968.

Smith, Lynwood. *Seashore Life of the Pacific Northwest*. Healdsburg, Calif.: Naturegraph Co., 1962.

Southward, Alan. *Life of the Seashore*. Cambridge, Mass.: Harvard University Press, 1965.

Stephens, William. *Southern Seashores: A World of Animals and Plants*. New York: Holiday House, 1968.

Index

(**Note:** A page number in boldface following an entry means that the entry is illustrated on that page. An asterisk following a page number means that a sound or sounds made by the entry are described on that page. A dagger following a page number means that a smell or smells emitted by the entry are described on that page.)